How to Use Your MULTITESTER

By
Alvis J. Evans

Electrical Inc.
An Applied Power Co.
Milwaukee, WI 53209

This book was developed and published by:
Master Publishing, Inc.
Richardson, Texas

Editor:
Gerald Luecke
Charles Battle

With contributions by:
Ed Filo

Design and artwork by:
Plunk Design

Printing by:
Arby Graphic Service, Inc.
Chicago, Illinios

Acknowledgements:
All photographs are either courtesy GB Electrical Inc. or Master Publishing, Inc.

9 8 7 6 5 4 3

Table of Contents

		Page
	Preface	iv
1	Getting Acquainted with a Multitester	1
2	Digital Multitesters	21
3	DC and AC Measurements	39
4	Measuring Individual Components	59
5	Measuring Components In Circuits	79
6	Home Appliance Measurements	97
7	Lighting and Related Systems Measurements	115
8	Automotive Measurements	129
9	Tool Control Circuit Measurements	143
	Appendix	151
	Glossary	152
	Index	155

Preface

Making basic electrical circuit measurements of voltage, current, or resistance with a meter helps significantly in understanding an electrical circuit to tell if it is operating properly or if repair or maintenance is necessary to correct its operation.

How to Use Your Multitester has been written to help you, the home-maker, "do-it-your-selfer," hobbyist, or technician, understand how to use multitesters to make basic electrical measurements. It very much emphasizes "how to do" the various measurements. In addition, it provides understanding of basic concepts and fundamentals, so that you have more appreciation of what is actually happening in the electrical circuit being measured. It is fully illustrated to visually enhance the understanding.

How to Use Your Multitester begins by showing you how to make practical, useful measurements needed around your home, your office, or your workshop. It shows how analog and digital multitesters differ, and the advantages and disadvantages of each. Following a general discussion about circuits and multitester measurements, where polarity, reference point, meter loading, accuracy, are the topics, basic dc and ac measurements are discussed with actual step-by-step procedures defined for voltage, current, and resistance measurements. These principles are then applied to measuring individual components — resistors, capacitors, inductors, transformers, semiconductor devices, batteries — individually and in circuits. Simple power supply, door-bell, and telephone circuits are used as examples. By making test measurements, you can determine the operation of a circuit.

How to Use Your Multitester continues by showing examples of common measurements that you can make using a multitester to determine a problem if something having an electrical system circuit is not operating properly. Chapters on home appliances systems, home power distribution systems, automotive systems, and power tool systems provide basic understanding and clearly illustrated examples. A few of the systems discussed include: heating and air conditioning, washers, dryers, low-voltage lighting, garage door openers, ceiling fans, alternators, batteries, ignition, and workshop power tools.

After reading this book and following its suggested procedures, you should be able to make circuit measurements and detect improper circuit operation. That was our goal; we hope we have succeeded — the rest is up to you.

AJE
MP

CHAPTER 1

Getting Acquainted with a Multitester

INTRODUCTION

Throughout this book, the word "meter" is encountered frequently. This word is a form of the French word "metre" which comes from the Greek word "metron," meaning to measure.

The British scientist Sir William Thomson, better known as Lord Kelvin (1824-1907), once simply and eloquently expressed the importance of measurement in the statement: "I often say that when you can measure what you are speaking about and can express it in numbers, you know something about it. When you cannot express it in numbers, your knowledge is of a meager and unsatisfactory kind."

We begin by examining the instruments used to measure the three parameters of Ohm's law — voltage (electromotive force or EMF), current and resistance. The meters used to measure these parameters are named for their respective unit — a voltmeter measures voltage in volts, an ammeter measures current in amperes, and an ohmmeter measures resistance in ohms. Often a single meter serves to measure more than one of these parameters, in which case it is called a *multimeter* or *multitester*.

Electricity seems elusive and mysterious. We know it is used to operate most of the equipment around the home and the office. Home lighting, appliances, heating and air conditioning, televisions, stereos, workshop tools, video games and home computers depend upon electricity for their operation. Although we can see and feel the effects of electricity, it is easier to understand the effects of electricity when it can be measured. That is the purpose of voltmeters, ammeters and ohmmeters.

This book has been written to help you understand how such multitesters work and how they can be used to make basic electrical measurements in the home, in the workshop, at the office, and on the job.

There are two basic types of multitesters used to make electrical measurements: analog and digital.

Analog Multitester

The analog multitester has a needle that deflects along a scale. The value of the measured parameter is indicated by the position of the needle on the scale. An analog meter is commonly called a voltmeter, ammeter, ohmmeter, VOM, multimeter or multitester depending on the parameters that it measures. The meter movement is generally of the D'Arsonval type. (We will explain this later.) Although the various analog multitesters may look different, all of them have scales, functions and ranges.

Figure 1-1 shows a typical analog meter. The main meter parts are: (A) the analog display and its meter movement (behind the scale plate — see *Figure 1-15)* which senses current and deflects its attached needle along the scale; (B) the switch(es) that set(s) the multitester's circuits to measure the chosen parameter and the full-scale range of the meter; and (C) adjustment controls and jacks that accept plug-in test leads which connect the meter to a component or circuit for measurement.

Digital Multitester

On a digital multitester, the measured value appears as a number on a digital display. It is commonly called a digital voltmeter, DVM, digital multimeter or digital multitester, DMM.

A typical digital meter is shown in *Figure 1-2.* The main parts are: (A) the digital display which indicates the numerical value of the measurement; (B) the switch(es) that set(s) the multitester's circuits to measure the chosen parameter and the full-scale range of the meter; (C) adjustment controls and jacks

Figure 1-1. Analog Meter

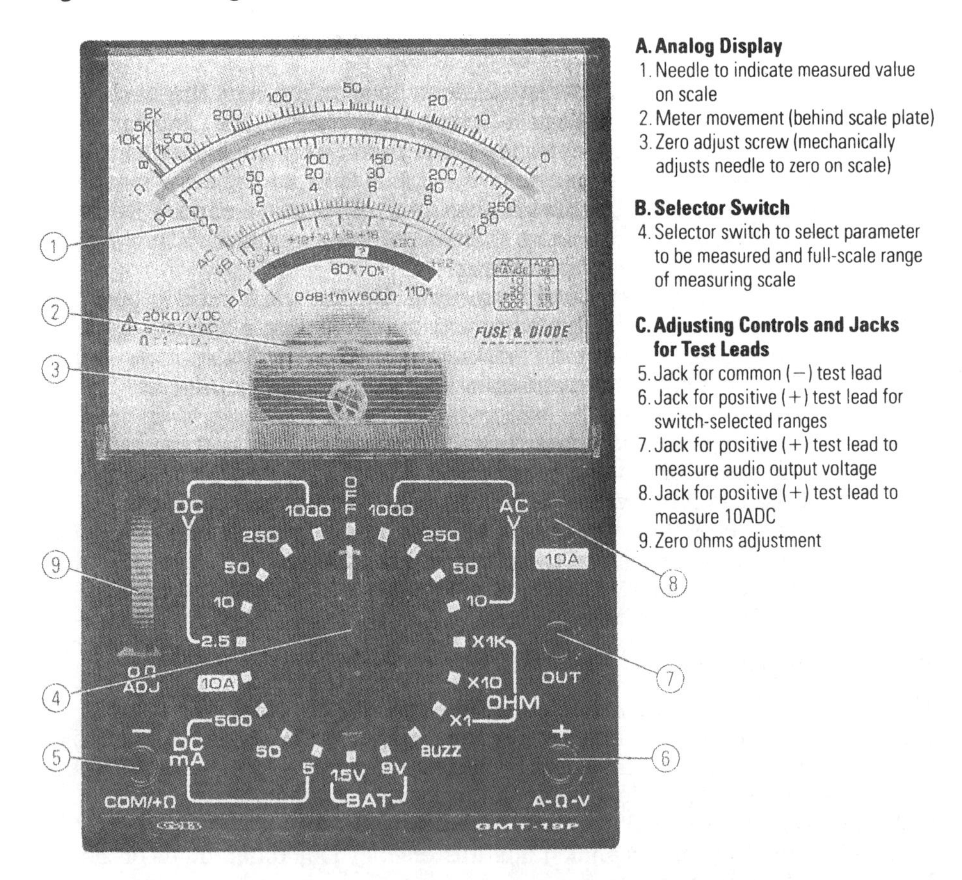

Figure 1-2. Digital Meter

A. Digital Display

1. Displays measured value in discrete digits from 0 to 999. Also displays polarity, overrange, low battery voltage, and if HOLD function is active.

B. Selector Switch

2. Selector switch to select parameter to be measured and full-scale range of measuring scale

C. Adjusting Controls and Jacks for Test Leads

3. Jack for common (−) test lead
4. Jack for positive (+) test lead for switch-selected ranges
5. Jack for positive (+) test lead to measure 10ADC
6. Push-button switch to activate the HOLD feature, which captures the measured value and retains reading in display when turned on.
7. Push-button switch to turn power on and off

that accept the plug-in test leads, and an ON-OFF switch which applies battery power to the internal circuits when in the ON position. A set of test leads with alligator clips to add to test leads is shown in *Figure 1-3.*

Let's first look at the most common type of meter — the analog volt-ohm-milliammeter (VOM).

GETTING TO KNOW YOUR ANALOG MULTITESTER

It would be inconvenient to carry around a collection of the various types of meters to make the different measurements of electrical and other physical parameters such as temperature. This was why the multitester was invented.

An analog multitester is shown in detail in *Figure 1-4.* Notice that the meter's scale plate has several calibrated scales which correspond to the different range selector switch settings. (We will examine the individual scales a little later.) The mirror on the scale plate is used to line up the needle and its reflection to improve the reading accuracy by preventing parallax error.

Figure 1-3. Test Leads

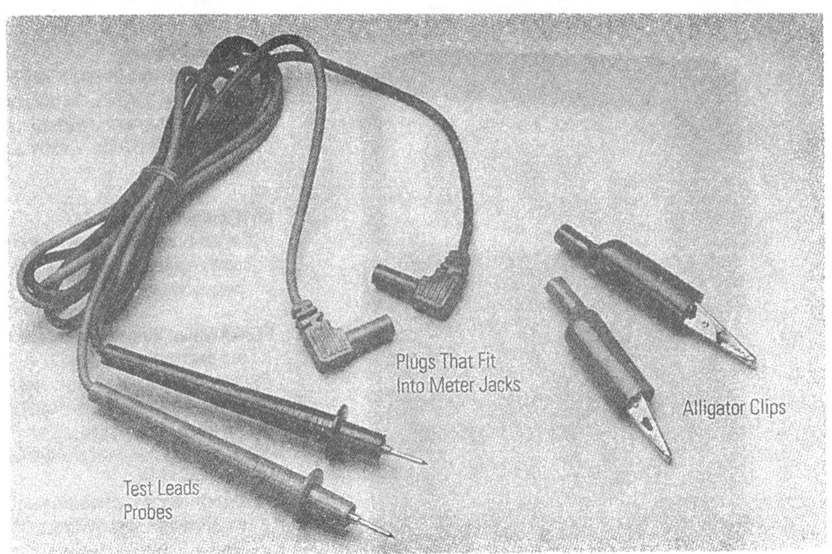

Refer again to *Figure 1-3* to examine the multitester test leads. One of the test leads is red and the other black. One end of each test lead has a probe for connecting to the circuit being measured and the other end has a plug designed to fit into a jack on the front of the multitester. The black lead is plugged into the jack marked "−COM" (which means "negative common"). The red lead is plugged into the jack marked "A-Ω-V" for most measurements. The red lead must be removed from this jack and plugged into the 10A jack for a 10A current range or the OUT jack to measure audio output voltages.

Another control on the multitester in *Figure 1-4* is the 0Ω ADJ (Zero Ohms Adjustment). This is used to keep the resistance measuring ranges calibrated as the internal battery ages. Its use will be described when we discuss making a resistance measurement a little later.

MEASURING A DC VOLTAGE

The most common measurement made with a multitester is EMF or voltage. There are two basic kinds of voltage as far as measuring with a meter is concerned. One kind has a force or pressure that is always in one direction (polarity) and is called dc (direct current) voltage. The other kind has a force or pressure that changes (alternates) direction (polarity) and is called ac (alternating current) voltage. Let's begin by measuring a 9-volt battery, a common dc voltage source.

A battery and most electronic equipment power supplies are sources of dc; therefore, they provide a fixed polarity voltage output. To measure their voltage, we must first locate the DCV (dc volts) section of the FUNCTION/RANGE selector switch. These positions are identified in *Figure 1-5b.* The ranges on this multitester are typical. The lowest range is 2.5V full scale and the highest switch-selected range is 1000V full scale. The additional ranges of 10V, 50V, and 250V are available by setting the FUNCTION/RANGE switch to

Figure 1-4. Analog Meter Details

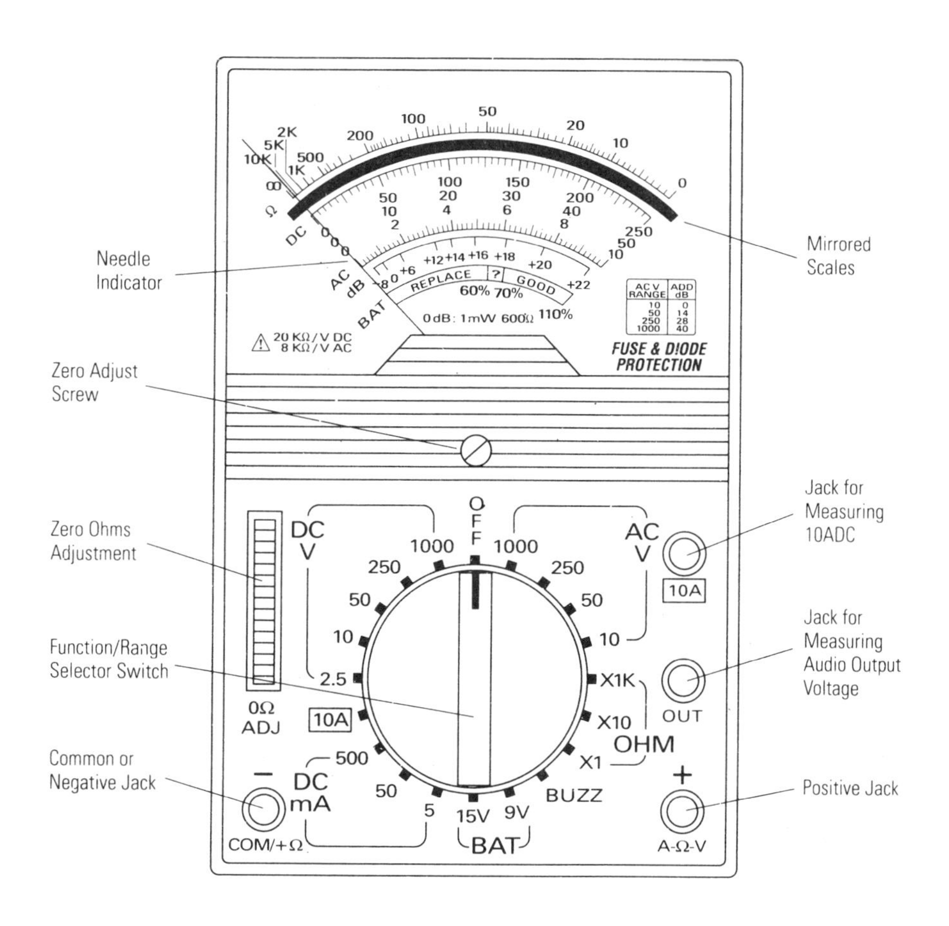

the appropriate position. To measure the 9-volt battery, select the 10DCV full-scale range as shown in *Figure 1-5b.*

With the test leads plugged into the proper jacks, touch the probe tips to the terminals of the 9-volt battery as shown in *Figure 1-5b.* Be sure that the polarity of the test leads is as shown — the red lead to the positive battery terminal and the black lead to the negative battery terminal. To measure the value correctly, you must identify the proper scale. Since the measured voltage is dc, find the three black scales that are marked "DC" as shown in *Figure 1-5a.* All three have 0 on the left-hand end and have either 10, 50, or 250 on the right-hand end. Also, on the left-hand end, you will see AC, dB, and BAT. These scales are used to measure ac voltage, decibels and a relative scale for battery condition, respectively. Since we are using the 10V full-scale range, we measure our voltage value on the scale that ends in 10 as shown in *Figure 1-5a.* The needle should be indicating approximately 9 volts, but *Figure 1-5a* shows 8.6V indicating the battery is partially discharged. Use the mirror behind the needle to "line up" the needle with its image for a more accurate reading. This

Figure 1-5. Measuring a DC Voltage (9V Battery)

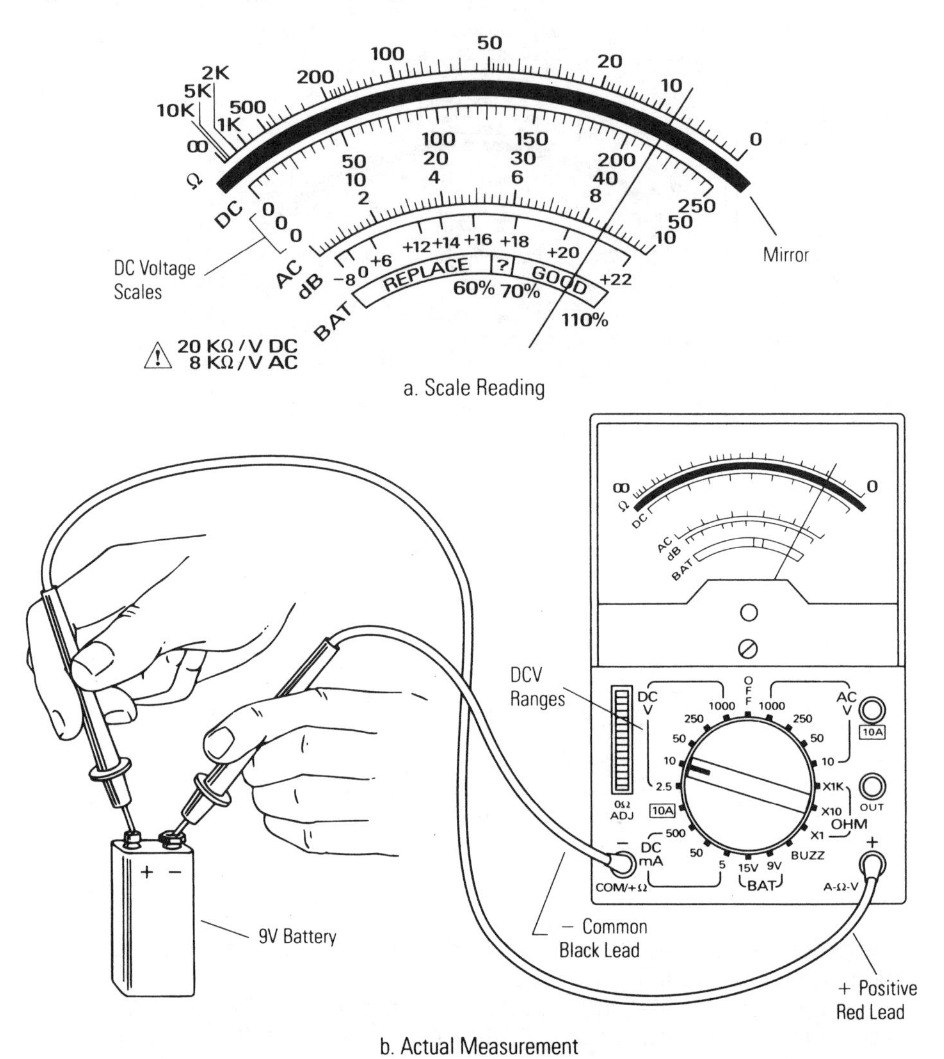

a. Scale Reading

b. Actual Measurement

makes sure your line-of-sight is perpendicular to the scale so you can accurately determine the position of the needle on the scale.

Before we go on to making an ac voltage measurement, many readers may be interested in more details about voltmeters. If you have no interest in this, you may skip ahead to the MEASURING AC VOLTS section.

BASIC DC VOLTMETER CONSTRUCTION

As we saw in *Figure 1-1*, the heart of any analog multitester is the D'Arsonval meter movement, which actually measures the average current through the movement. Because the current which causes the deflection of the meter movement is related by Ohm's law to the applied voltage (EMF) across the circuit and the circuit's resistance, we can use the meter movement to indicate any of the three parameters: current, voltage or resistance. In the application above, we used the dc voltmeter function to measure the voltage of a 9-volt battery.

A simplified circuit (schematic) diagram of a basic dc voltmeter is shown in *Figure 1-6.* The value of the multiplier resistor, R_X, determines the amount of voltage required to move the meter movement's needle to the full-scale position. On all of the voltmeter ranges shown in *Figure 1-5,* a multiplier resistor is placed in series with the meter movement by the FUNCTION/RANGE Selector switch.

When a voltage E appears across the multitester probes, there is a current, I, in the series circuit of *Figure 1-6.* When E is large enough to cause a full-scale deflection on the meter, the current is identified as I_M and is called the meter sensitivity — the current required to deflect the meter to full scale. The voltage, V_M, is the amount of voltage appearing across the meter terminals due to the resistance, R_M, of the meter. Knowing I_M, and measuring V_M, the resistance R_M of the meter movement can be determined. For the meter of *Figure 1-5,* I_M is 50 μA and R_M is equal to 5,000 ohms. Multiplier resistors, R_X, are added to the circuit of *Figure 1-6* so that the meter movement can be a dc voltmeter with multiple ranges to measure voltages over a wide range of values without exceeding I_M on any range. For the full-scale voltage, E, applied for each range, $R_X + R_M$ must limit the current to I_M.

Refer again to the basic circuit shown in *Figure 1-6.* The value of the multiplier resistor, R_X, determines the voltage range for full-scale deflection. For example, using Ohm's law R = E/I, the 10V range requires a total resistance of 200,000Ω to limit the current to 50μA full scale (10V/50μA). $R_X + R_M$ equals the

Figure 1-6. The Basic Circuit for a DC Voltmeter

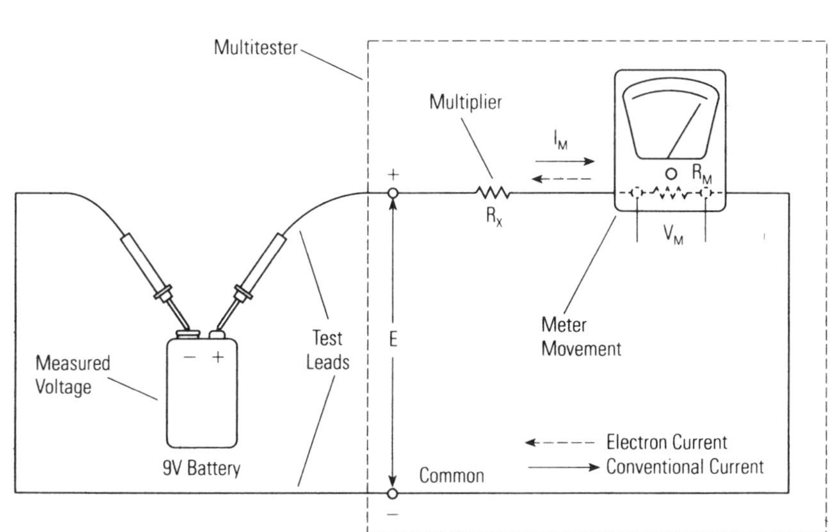

total resistance. R_M, the meter movement resistance is 5000Ω, so R_X, the multiplier resistor for the 10V range, is 200,000Ω − 5000Ω = 195,000Ω. The multiplier resistor for any range may be found by this method. Simply divide the voltage range by 50μA to find the total resistance, and then subtract the meter movement's resistance to obtain the value of R_x.

Multirange Voltmeters

The selection of one of a number of multiplier resistors by means of a range switch provides a voltmeter with a number of voltage ranges. *Figure 1-7* shows a multirange voltmeter using a four-position switch and four multiplier resistors. The multiplier in each range was calculated using the simple method for R_X given previously.

A variation of the circuit of *Figure 1-7* is shown in *Figure 1-8*. In this circuit, V_4 is a higher voltage range than V_3, V_3 is higher than V_2, and V_2 is higher than V_1. As a result, the multiplier resistor of a lower range can be included as part of the multiplier resistance for each higher range. An advantage of this circuit is that all multiplier resistors except the one for the lowest range have standard resistance values which are less expensive and can be obtained commercially in precision tolerances. The first multiplier resistor, R_1, is the only one that must be specially manufactured to meet the particular requirements of the meter movement.

Voltmeter Sensitivity

The value of the voltmeter resistance (referred to as input resistance) depends on the sensitivity of the meter movement. The greater the sensitivity of the meter movement, the smaller the current required for full-scale deflection and the higher the input resistance. The term "ohms per volt" is used exclusively to refer to voltmeter sensitivity. For the meter movement that takes 50μA for full-

Figure 1-7. Simple Multirange DC Voltmeter

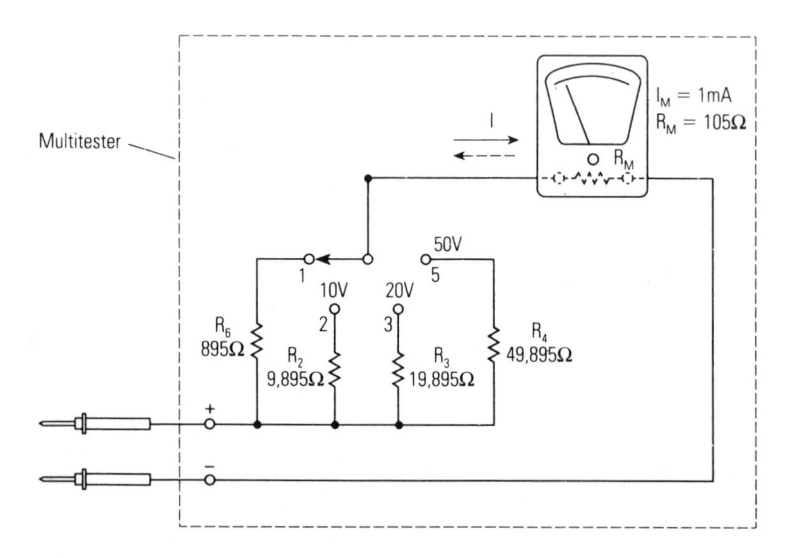

Figure 1-8. Alternate Multirange DC Voltmeter Circuit

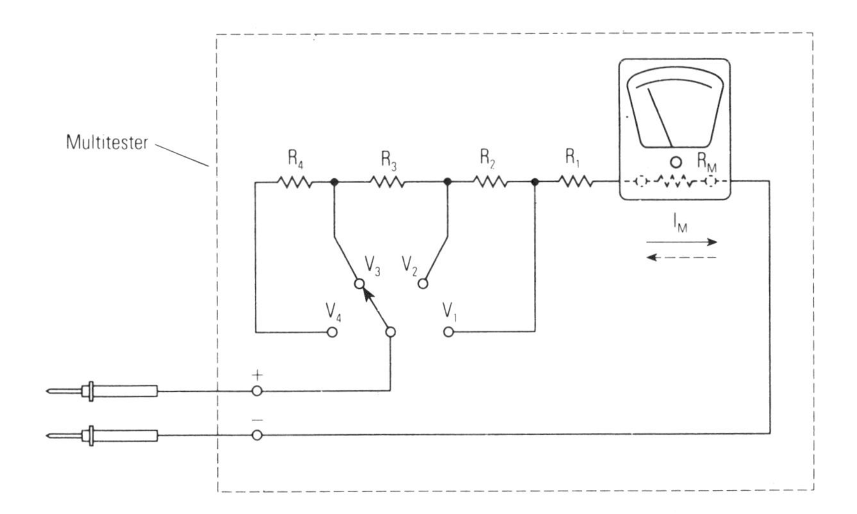

scale deflection, the meter sensitivity is 20,000 ohms per volt. If the full-scale voltage range is set on the 1-volt range, the input resistance is 20,000 × 1 or 20,000 ohms; if it is set on the 25-volt range, the input resistance is 20,000 × 25 or 500,000 ohms. The multitester of *Figure 1-4* is a 20,000Ω/V voltmeter.

MEASURING AC VOLTS

The electric power generated by the utility companies is ac (alternating current) which alternates at a frequency of 60Hz. The power is distributed over high-voltage lines, but before it enters a customer's house, transformers reduce the voltage to 230 volts for high-current appliances such as ovens, and to 115 volts for lights and other appliances. The ac power is distributed throughout a home, workshop, office or business through wiring connected to ac wall outlets. Let's see how we can measure the ac voltage at one of these wall outlets.

WARNING

The ac line voltage from a wall outlet is very dangerous. Always hold the test probe far back from the metal tip. *Never touch the metal part of the probes when measuring voltage with the multitester.* Wear rubber-soled shoes, if possible, and do not stand in water or on damp ground or floor when making these measurements. If possible, use only one hand to make the measurement. The old rule, "keep one hand in your pocket," is still a good one.

The diagram for measuring the voltage at a wall outlet is shown in *Figure 1-9b.* To measure an ac voltage, we must set the FUNCTION/RANGE selector switch to one of the AC voltage positions. For the multitester used in *Figure 1-9*, the ac voltage measuring function is labeled ACV. Set the FUNCTION/RANGE selector switch to the 250ACV range as shown in *Figure 1-9b.*

Figure 1-9. Measuring an AC Voltage (115VAC Outlet)

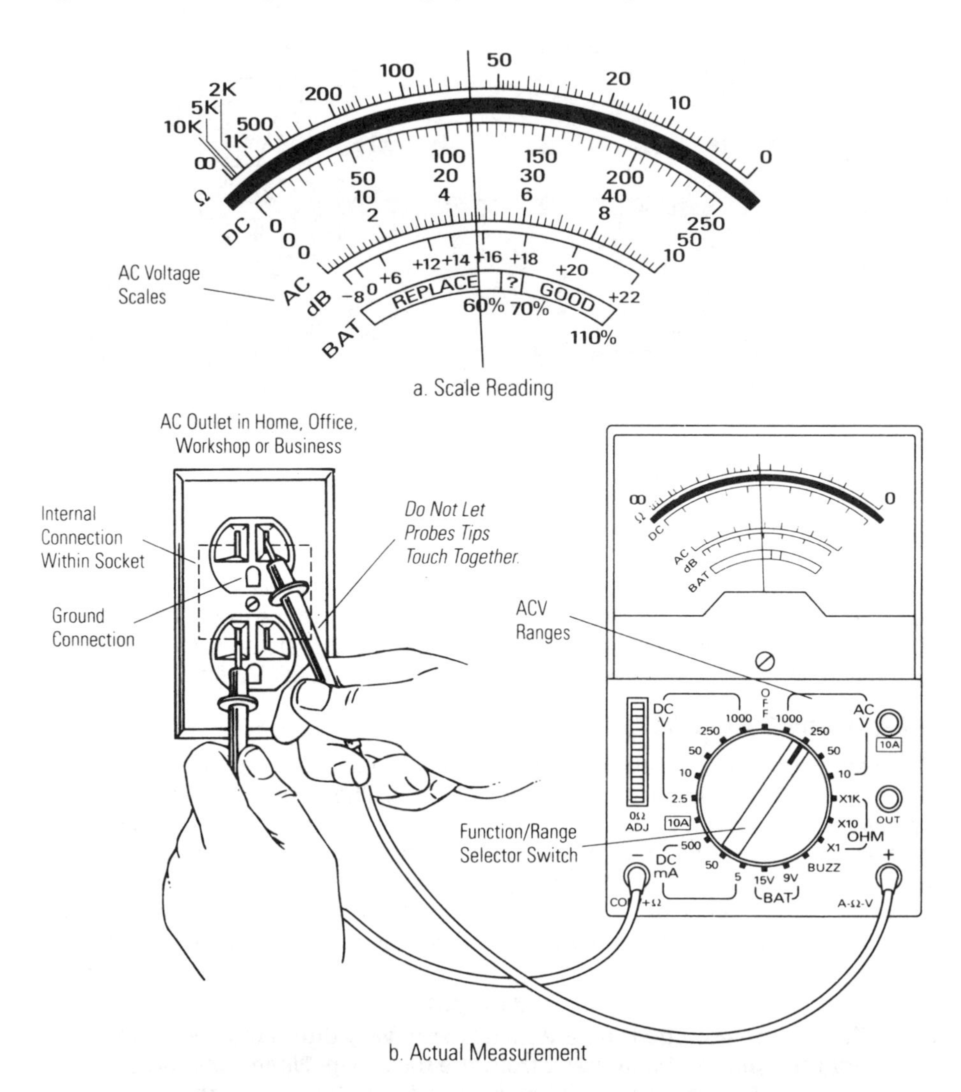

Plug the black test lead into the −COM jack and the red test lead into the A-Ω-V jack as you do for dc measurements. Actually, the connection polarity does not matter in an ac voltage measurement, but it is good to be consistent. Be sure to observe the warning above.

Place the test lead probes in the ac outlet socket connections as shown in *Figure 1-9b.* There are internal connections within the outlet so that each test lead can measure one terminal on each socket and have enough space between them so that you will not short out the test leads. Do not put both test leads on one socket. If you do, you might short out the test leads and cause sparking and damage to the test leads and the socket.

Read the voltage value on the red scales marked AC or ACV as shown in *Figure 1-9a*. Use the same black numbers that were mentioned above in the dc voltage measurement discussion. Read the value on the red scale using the black number set which ends in 250. The voltage should read between 115 and 120 volts as shown in *Figure 1-9a*. Use the scale mirror to align the needle to eliminate parallax error.

BASIC AC VOLTMETER CONSTRUCTION

The most practical means for measuring commonly used ac voltages is by combining a rectifier with the highly sensitive D'Arsonval meter movement. The rectifier changes the alternating current to direct current. (The D'Arsonval meter movement requires dc because a permanent magnet provides the magnetic field for the moving coil. See *Figure 1-15.)*

The schematic diagram of a basic ac voltmeter circuit is shown in *Figure 1-10*. Notice that the D'Arsonval meter movement is used just as in the dc voltmeter discussed above. However, as mentioned for the outlet voltage measurement, you do not have to observe polarity because the rectifier takes care of sending the current through the meter movement in the correct direction. The correct multiplier resistor for the particular range is connected by the range selector switch. If the measured voltage is less than the full-scale value of the next lower range, then you may switch to the next lower range to get a more accurate reading.

Bridge-Type Rectifier

The circuit of *Figure 1-10* makes use of a bridge-type rectifier which provides full-wave rectification. The bridge is generally made of germanium or silicon diodes. A D'Arsonval meter movement provides a deflection that is proportional to the "average" value of the DC current.

Figure 1-10. AC Voltmeter Using a Bridge Rectifier

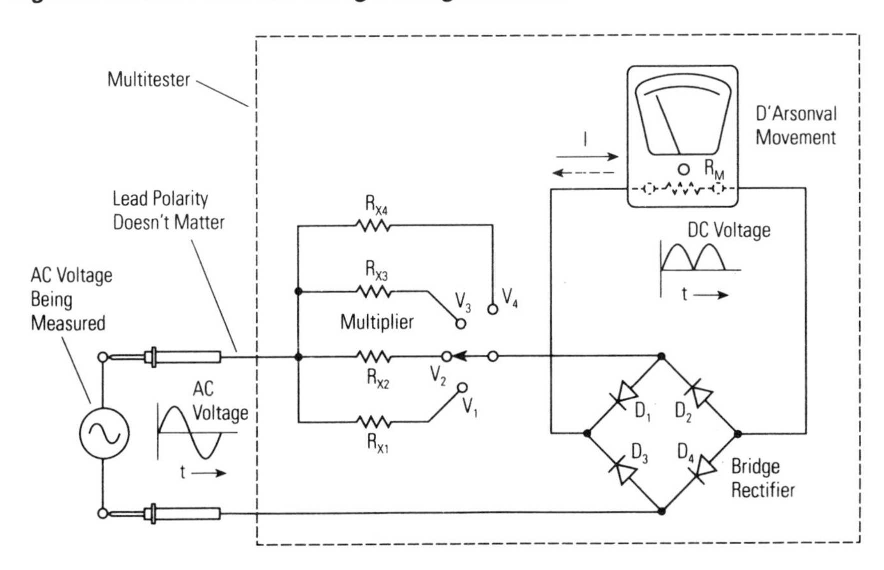

MEASURING DC CURRENT

Most multitesters can measure dc current but not ac current. The multitester shown in *Figure 1-4* has four dc current ranges — from 5mA to 500mA (0.5A), or 10A with special jack. *Figure 1-11* shows an example of using the multitester to measure the current drawn by a miniature 6-volt lamp. This is a simple series circuit with a battery, lamp and meter in the circuit. The ammeter measures current through an element of a circuit only when it is inserted into the circuit in series with the particular element. WARNING: *Never connect an ammeter **across** the circuit element* — damage to the multitester will usually result. It does not matter where in a simple series circuit that the ammeter is placed since the current is the same everywhere in a series circuit.

To measure a dc current, we must set the FUNCTION/RANGE selector switch to one of the dc current positions, so set the FUNCTION/RANGE selector to the 500mA range. Connect one terminal of the lamp to the negative battery terminal. Plug the black test lead into the −COM jack and the red test lead in the A-Ω-V jack. Touch the red (positive) test lead of the meter to the positive post of the battery and touch the black (negative) test lead to the other lamp terminal. Read the current on the DC (black) scale that ends in 50, but move the decimal point one place to the right (the 50 is actually 500). If the current reads less than 5 (50mA), switch the FUNCTION/RANGE selector to the next lower range for a more accurate reading. Remember to move your head to align the needle and its mirror image so you obtain the most accurate reading. The reading in *Figure 1-11a* is 150mA.

The next section, CONSTRUCTION OF AN AMMETER, presents more detail on ammeters. If you are not interested in this detail, you may skip ahead to the BASIC OHMMETERS section.

CONSTRUCTION OF AN AMMETER

The simplest ammeter contains only the basic meter movement. It can measure values of current up to its sensitivity rating (full-scale deflection). Meter movements that require a large current for full-scale deflection are not practical because the large wire they would require on the coil would have a large mass. Therefore, highly sensitive meter movements are desensitized by placing low value resistors called "shunts" in parallel with the meter movement to extend their current handling range. If a meter movement requires 100 microamperes for full-scale deflection and the meter is to have a 10-milliampere (10,000 microamperes) scale, the shunt must have a resistance value so that it will bypass 9900 microamperes around the meter movement.

Simple Ammeters

Figure 1-12a shows a single 20.202Ω shunt across a basic 100μA meter movement to extend the full-scale range of the movement to 10mA. Shunts are easily calculated by applying Ohm's Law (E = IR) and the principles of a parallel circuit, if the meter movement's internal resistance R_M is known.

Multirange Ammeters

What we have said is that a basic meter movement has a given sensitivity — a given amount of current that must pass through it to produce full-scale deflec-

Figure 1-11. Measuring the Current Drawn by a Lamp

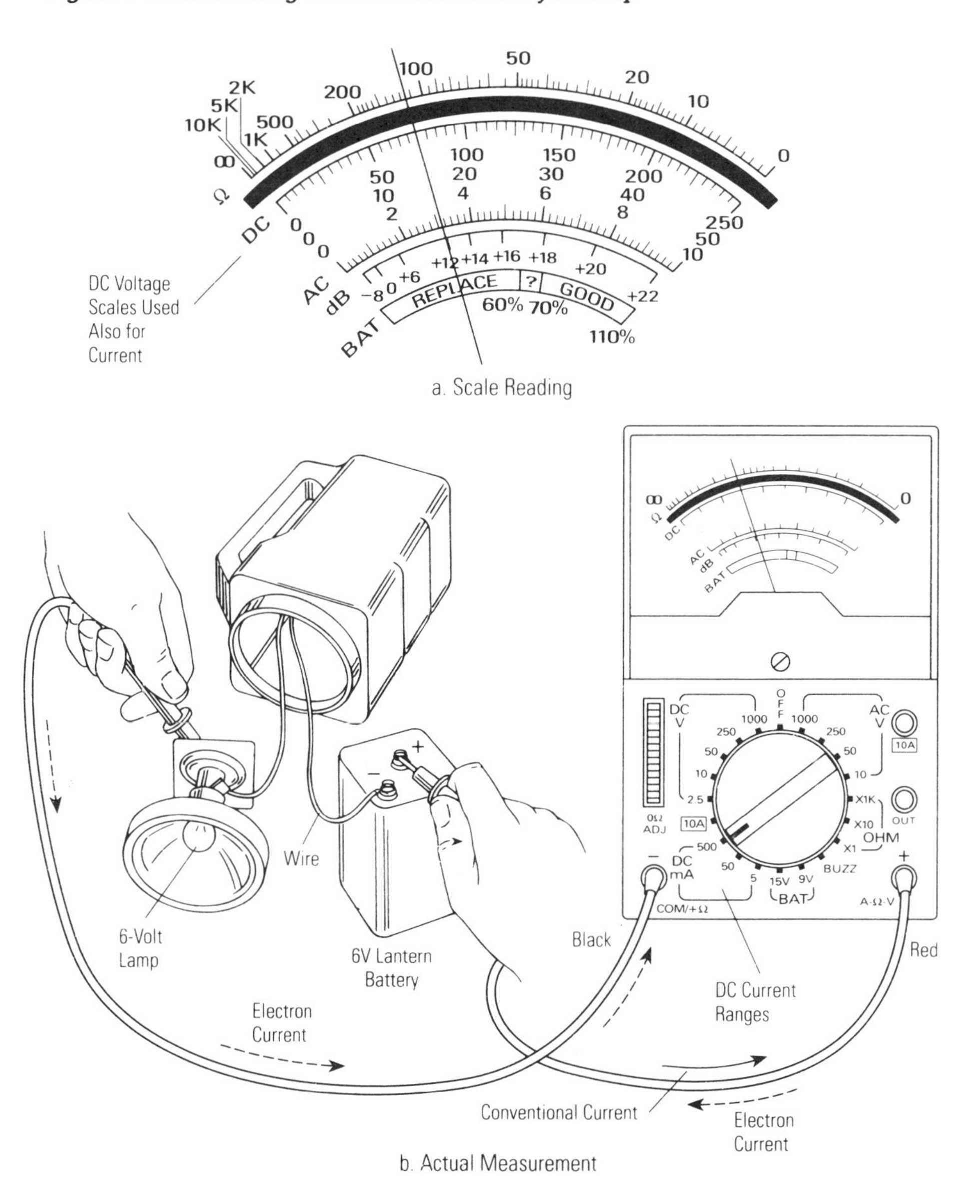

tion. By using suitable shunts, the range of current that a given movement can measure is increased. This is essentially the process used in the construction of multirange ammeters. In *Figure 1-12b*, a four-range ammeter is shown. The meter leads are used to connect the meter in the circuit to measure a current. Note the lowest range (position 1) uses the meter's basic sensitivity of 1 milliampere (1 mA) with no shunt. The values of the shunts are given for the other ranges. These values may be verified as for *Figure 1-12a*. The fifth

Figure 1-12. Ammeter Schematics

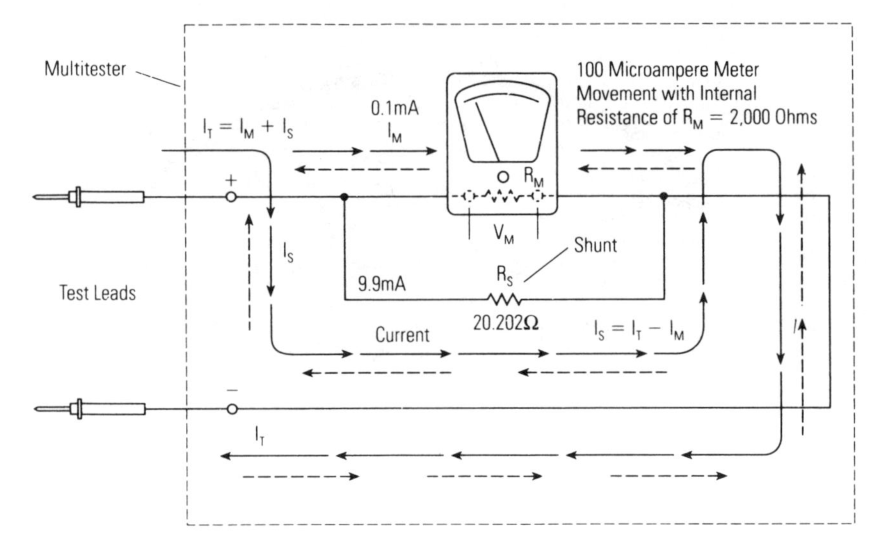

a. Simple Ammeter-Single Shunt

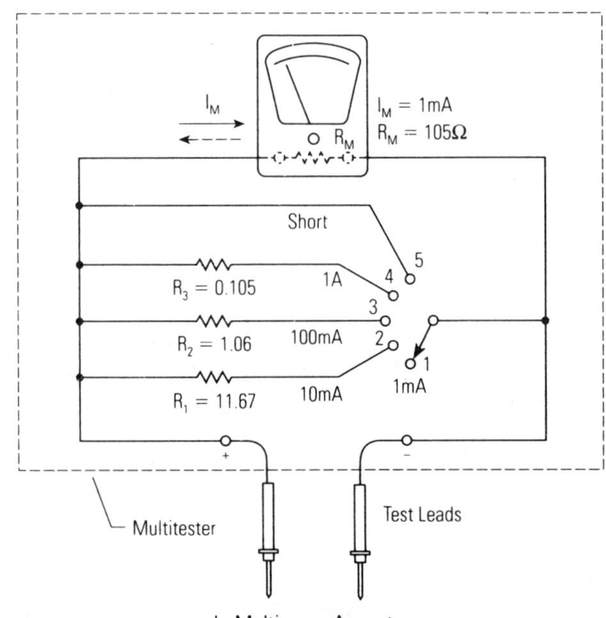

b. Multirange Ammeter

position on the switch is marked SHORT which takes the meter out of the circuit without physically disconnecting its leads. This position is also recommended when transporting a meter. It provides further damping to prevent damage to the meter.

BASIC OHMMETERS

Since electrical resistance is measured in ohms, a common instrument used to measure resistance is called an ohmmeter. Recall that according to Ohm's law, the resistance, R, in a dc circuit can be determined by measuring the current, I, in the circuit as a result of applying a voltage, E. The voltage divided by the current is the resistance value, or

$$R = \frac{E}{I}$$

The resistance is one ohm if there is one ampere of current in a circuit when one volt is applied. By applying a known voltage, E, to a circuit and measuring the current, I, the meter scale can be calibrated to indicate the resistance in ohms. This is the basic principle of an ohmmeter. An ohmmeter is actually measuring the current in a circuit when a known voltage is applied to the circuit. The known voltage is contained in the multitester. The ohmmeter scale is marked off in ohms rather than units of current.

Figure 1-13 shows a pictorial view of an ohmmeter that includes the internal circuit. Small subscripts are placed on the resistance symbol (R_V, R_C, R_M, R_K,) to identify the different resistors in the circuit. The unknown resistance, R_U, is measured by placing the test leads across it as shown in the figure. The internal battery applies an EMF, E, of a given voltage, V, to the circuit consisting of the Zero Ohms Adjustment resistance, R_C; the meter movement internal resistance, R_M; a known scaling resistance, R_K; and the unknown resistance, R_U. When $R_U = 0$, there is full-scale meter movement current in the circuit, and the needle deflects to the 0 (zero) ohms position on the right-hand side of the meter scale. When R_U is an open circuit ($R_U = \infty$), then the needle does not move and indicates infinity on the ohms scale because there is no circuit current.

MEASURING RESISTANCE

Measuring resistance with the ohmmeter function of a multitester is somewhat different from measuring voltage or current. Never attempt to measure resistance of a wire, circuit or other device if it has power applied.

CAUTION:

Always remove the source of power to a device or circuit before you attempt to measure resistance.

Figure 1-13 shows how to measure a resistance. When we discussed the multitester in *Figure 1-4*, we mentioned the 0Ω ADJ (Zero Ohms Adjustment) control. For an accurate reading of the resistance of a component or device it is absolutely necessary to zero the ohmmeter first. This is done by setting the FUNCTION/RANGE switch to one of the resistance measuring ranges as shown in *Figure 1-13*. Plug the black test lead into the −COM/+Ω jack and the red test lead into the A-Ω-V jack. Hold the ends of the black and red test leads together while you adjust the 0Ω ADJ until the needle is exactly over the 0 on the OHMS scale. Disconnect the meter test leads from one another and place them across the unknown resistance. When you switch to another ohmmeter range, check the zero adjustment again. If you cannot adjust the needle to 0 on the OHMS scale, the most likely cause is low battery voltage. It is good practice to replace the internal battery (or batteries) once a year.

Figure 1-13. Measuring Resistance of a Resistor with an Ohmmeter

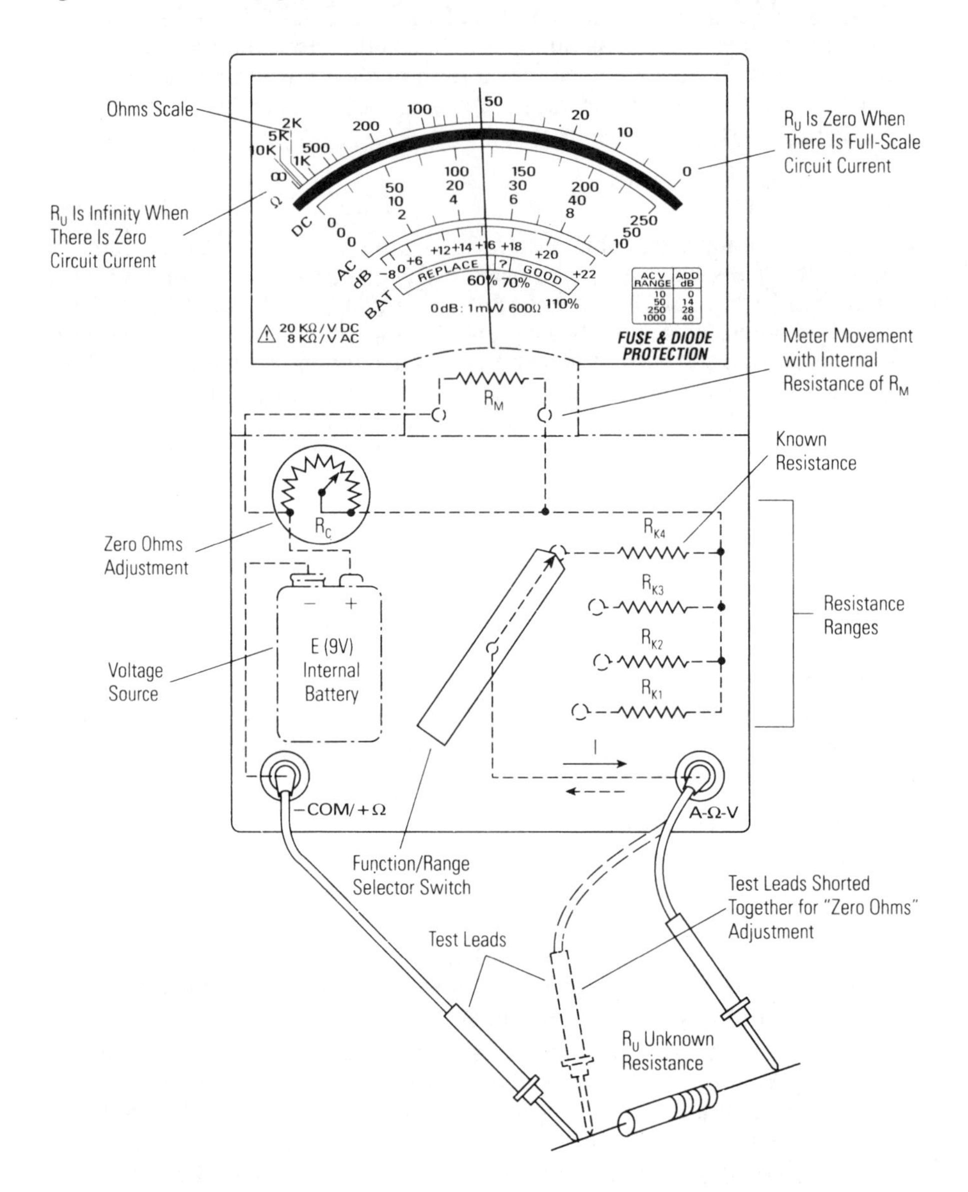

MEASURING CONTINUITY

When all the components and interconnecting wires in a circuit are connected properly, we say there is good continuity in the circuit. If any wire is not making a good connection, it will not have continuity. When we measure the continuity of a wire or connection, we are essentially using an ohmmeter to measure for $R_U = 0$.

When measuring continuity, the actual value of resistance usually is not of interest, but only whether or not there is continuity; that is, whether there is a complete electrical circuit. An example of measuring the continuity of a typical power cord for a deep fryer with an ohmmeter is shown in *Figure 1-14.* To test for continuity, set the FUNCTION/RANGE Switch to the X1 range and zero the ohmmeter as described above. Connect either one of the test leads to a plug prong on one end of the power cord and the other test lead to one of the connector jacks (a hole) on the other end of the power cord. If there is continuity, the meter will deflect to zero ohms. If continuity is not indicated, move the meter lead to the other hole because continuity will be indicated only when you are testing both ends of the same wire. If continuity still is not indicated, then the power cord is defective. If continuity is indicated, move both test probes to test the other wire in the same way. While the test probes are connected in each test condition, flex and stretch the cord, especially near the connectors, to test for intermittent continuity.

OHMMETER USE

Ohmmeters are very useful in servicing electronic equipment by making quick measurements of resistance values. The resistance values that can be measured with the ohmmeter vary from a fraction of an ohm to 100 megohms or more. The measurement accuracy is usually about 3%, which is adequate for most troubleshooting of electrical circuits. However, this accuracy is not suitable for precision laboratory measurements.

Since ohmmeters contain their own power source, and depend on their internal calibrated circuits for accuracy, they should be used only on passive circuits; that is, circuits that are not connected to power sources. *Connecting an ohmmeter into a "live" circuit will almost surely destroy the calibration and can very easily destroy the ohmmeter.*

Series Ohmmeter Circuit Details

For readers that are interested, let's look again at the basic series-type ohmmeter circuit shown in *Figure 1-13.* As mentioned previously, when the test leads are open ($R_U = \infty$), there is no current through the meter. The meter needle remains at rest to indicate an infinite resistance.

When the test leads are shorted ($R_U = 0$), there is a full-scale deflection of the needle indicating a zero resistance measurement. A typical meter scale is shown in *Figure 1-13.* The purpose of R_C, the Zero Ohms Adjustment control, is to allow adjustment to compensate for a changing battery potential due to aging, and for lead and fuse resistance. The purpose of R_K is to limit (and set) the current through the meter circuit to full scale when the test leads are shorted, therefore, indicating zero ohms on the scale.

It is convenient to use a value for R_K in the design of the series ohmmeter such that when an unknown resistance equal to R_K is measured, the meter will deflect to half scale. Therefore, when the leads are shorted together, the meter deflection will be full scale. Using this criteria, multiple ranges for the ohmmeter can be provided by switching to different values of R_K as shown in *Figure 1-13.* Each R_K sets the half-scale value for the respective resistance range.

Figure 1-14. Measuring Continuity of a Power Cord with an Ohmmeter

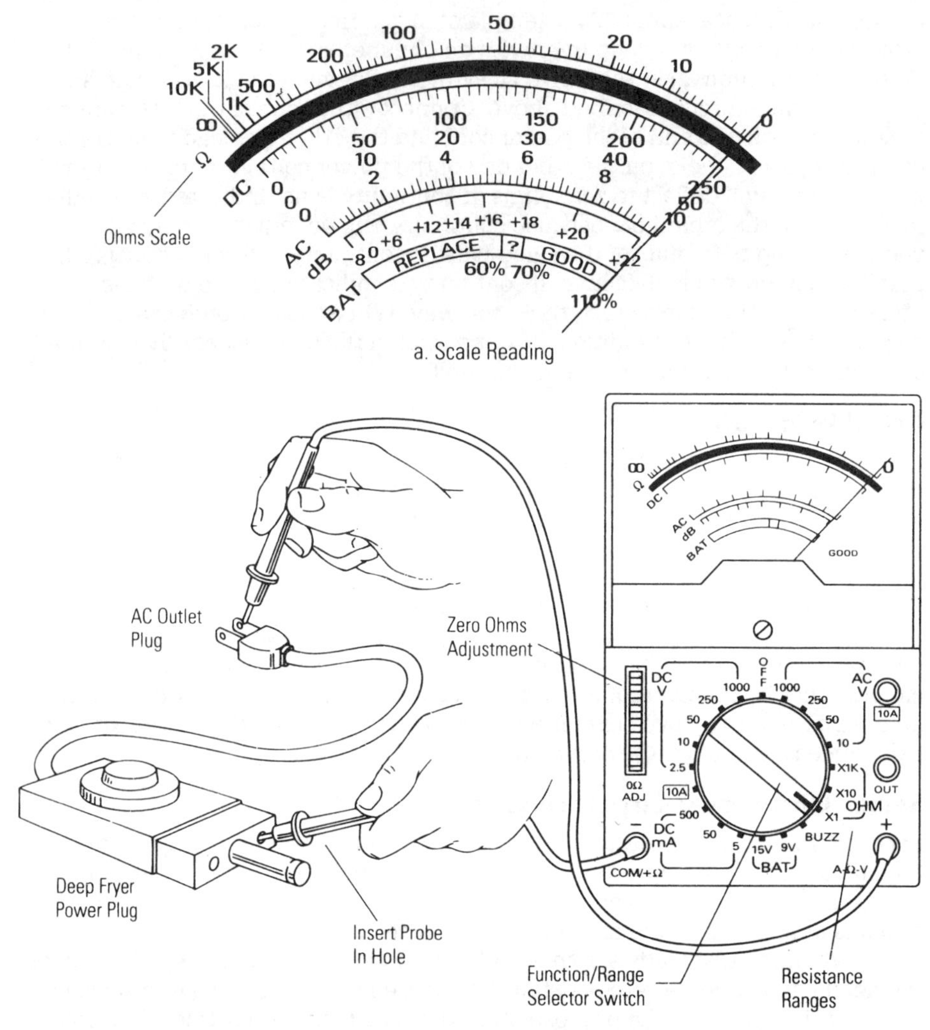

OPERATING HINTS

Having covered some of the types of analog meters, let's review procedure, accuracy, parallax, battery replacement and protection — items that apply generally.

Procedure

Before making a measurement, set the FUNCTION/RANGE selector switch to the proper function and range. If you do not know the value of the parameter you are about to measure, set the FUNCTION/RANGE switch to the highest range. Momentarily touch the test leads to the measuring points while observing the needle. If the needle deflects very quickly upscale and hits the right-hand stop, disconnect the test leads immediately. This parameter exceeds the full-scale capability of the multitester and you should not attempt to measure this parameter with this multitester. If the reading is low on the scale, you may switch to a lower range.

You must observe polarity in dc measurements. To determine if the polarity of the test lead connections is correct, momentarily touch the test leads to the measuring points while observing the needle. If it deflects upscale, the polarity is correct and you may proceed with the measurement. If it deflects downscale and goes below zero, the polarity is not correct. Since you caused only momentary current, the meter should not be damaged. Reverse the test lead connections and try again.

Accuracy

Multitester movements and circuits have basic accuracies. For analog multitesters, the typical accuracy of the D'Arsonval movement is 3-4%. The stated accuracy in percent refers to the percent of full-scale reading on any range. For less meter error, always select the range where the reading is nearest full scale. See *Figure 1-15*.

Parallax

A mirror is provided on the scale of some meters to help you eliminate parallax error. This error is caused when your line-of-sight is not perpendicular to the meter scale when you take the reading. To eliminate the parallax error, move your head so that the needle is exactly over its image in the mirror. If the meter does not have a mirror, try to accomplish the perpendicular line-of-sight anyway.

Weak Batteries

Internal batteries need to be replaced if the Zero Ohms Adjustment control no longer adjusts the needle to R = 0. The amount of adjustment range is a good indicator of battery condition. As mentioned previously, it is a good idea to replace the batteries once a year to maintain the multitester in top operating condition.

Diode Protection

To protect the movement from overload, most multitesters have two diodes connected in parallel "back-to-back" across the meter movement. Therefore, the voltage across the meter movement cannot exceed about 0.7 of a volt. If these diodes have been damaged by a measurement condition, it is almost certain that the meter movement has been damaged.

Fuse Protection

A fuse is a current-sensitive device that is designed to melt and produce an open circuit when the current exceeds a predetermined level. If the multitester does not operate on any function or range (attempting a zero ohms adjustment is a quick check), it is likely that the protection fuse has blown. This fuse is usually located inside the case. Follow your operator's manual to replace the fuse and the multitester should be operable again.

Figure 1-15. D'Arsonval Moving Coil Meter Movement

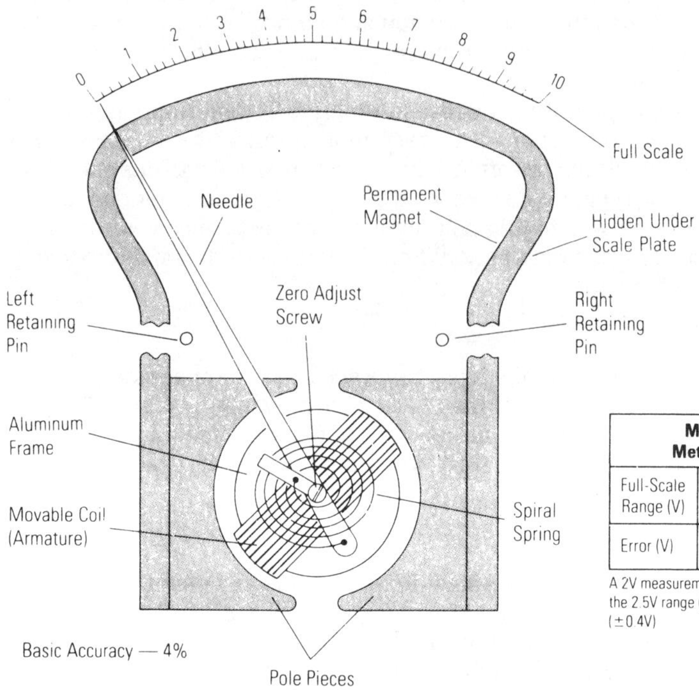

Measurement Error Meter Accuracy — 4%			
Full-Scale Range (V)	2.5	5	10
Error (V)	±0.1	±0.2	±0.4

A 2V measurement will be more accurate on the 2.5V range (±0.1V) than on the 10V range (±0.4V)

Operation of the Basic Meter Movement

Let's look further at the basic construction and operation of the D'Arsonval meter movement *(Figure 1-15)*. The movable coil with many turns, called an armature, has a needle attached to it. The armature is located within the strong magnetic field of the permanent magnet (PM) which is behind the scale plate. The uniform radial field surrounding the moving coil is required to make the torque produced by the current in the coil result in a linear movement of the meter needle along the calibrated scale. As current increases, the deflection increases.

One end of the needle is attached to a spiral spring so the needle will be returned to an initial position when the current in the coil is removed. The other end of the spiral spring is attached to a ZERO ADJUST screw located on the core of the moving coil. By turning the ZERO ADJUST screw with no current applied, you adjust the needle to its static or mechanical zero position. If current passes through the meter movement coil in the wrong direction, the needle deflects backwards against the meter's left-hand retaining pin. This is one of the most common ways of damaging or burning out a meter — *passing too much current through the coil in the wrong direction.*

SUMMARY

Now that we know how to use analog multitesters to measure current, voltage and resistance and some of their characteristics, let's look at digital multitesters.

CHAPTER 2

Digital Multitesters

BASIC DMM

A typical digital multitester or multimeter (DMM) was shown in *Figure 1-2.* It is shown again in *Figure 2-1.* It has a single rotary function/range selector switch that not only selects whether the meter is used as a dc voltmeter, dc ammeter, ac voltmeter, or an ohmmeter, but selects the full-scale range of the measurement as well. It has a digital display, an ON-OFF switch for POWER and HOLD, and test jacks to accept test leads. It has 14 ranges. The digital display shows the value and polarity of the quantity measured. It indicates when the internal battery voltage is low, and when the HOLD function is active. The measured value is held in the display when the HOLD switch is ON.

Comparison to Analog Meters

What does "analog" mean and what does "digital" mean? Let's look at *Figure 2-2.* As shown, an analog signal varies continuously in a smooth fashion without any breaks in the signal. If an analog meter were measuring the signal, the meter needle would vary continuously as the signal varies.

A digital meter, on the other hand, would sample the analog signal at regular intervals as shown at A, and would convert the value of the analog signal at the sampled time into a digital code made up of separate parts, or bits. The digital codes for the sampled values are shown at B. The digital codes can be processed in the meter with logic and decision circuits, and converted to signals that drive the meter's digital display.

The digital value changes only at the sampled time, and the displayed value is represented in digits the same way that numbers are displayed on a calculator as shown at C. Therefore, the DMM measures analog quantities (voltage, current, temperature, etc.) and displays them in digital (discrete) numerical values.

The digital multimeter of *Figure 2-1* has a digital display, wherein the measurement value is displayed as a four-digit number. On the lowest full-scale voltage range, the measurement is read on the display to an accuracy of three decimal places.

Unlike the analog meter, digital multimeters contain electronic circuits to produce their measurement value rather than an electromechanical meter movement. Because they are designed to be completely portable, they need internal batteries to supply power for the electronic circuits as well as an energy source to supply current for resistance measurements when the meter is used as an ohmmeter. When turned ON, the ON-OFF switch connects the power source to the appropriate circuits. Like the analog multitester, most

Figure 2-1. Digital Multimeter

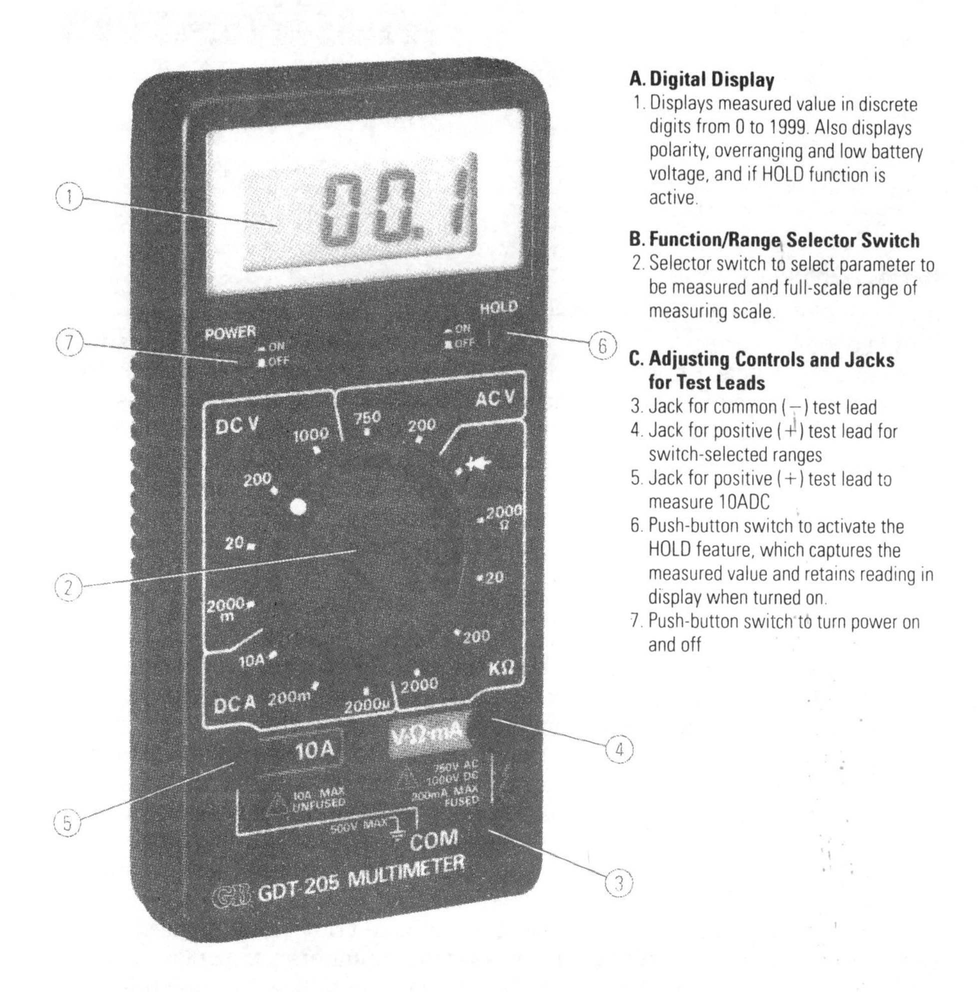

A. Digital Display

1. Displays measured value in discrete digits from 0 to 1999. Also displays polarity, overranging and low battery voltage, and if HOLD function is active.

B. Function/Range Selector Switch

2. Selector switch to select parameter to be measured and full-scale range of measuring scale.

C. Adjusting Controls and Jacks for Test Leads

3. Jack for common (−) test lead
4. Jack for positive (+) test lead for switch-selected ranges
5. Jack for positive (+) test lead to measure 10ADC
6. Push-button switch to activate the HOLD feature, which captures the measured value and retains reading in display when turned on.
7. Push-button switch to turn power on and off

digital multitesters have one switch that selects both the function and the full-scale range of the measurement; however, some have two separate switches, as well as jacks or sockets for special functions. All have jacks that accept test leads.

The digital display for the meter of *Figure 2-1* has three full digits. Each of them can be a numeral from 0 through 9. In addition, there is a digit on the far left, called the most significant digit, which can be only the numeral 1. It will be turned on, as the measured value increases, along with the other three digits, until the measured value reaches full-scale of the range selected. The most significant 1 will be on by itself when the measured value exceeds the range selected. A DMM with this type of display is referred to as a 3-1/2-digit DMM — more on this later.

Figure 2-2. Analog vs. Digital

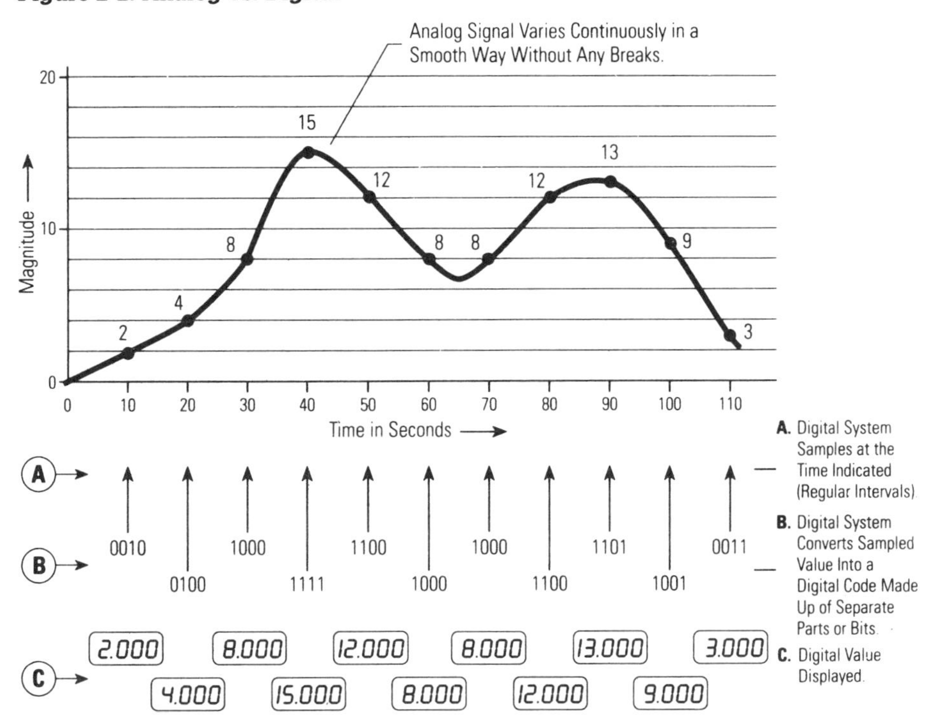

DMM Characteristics

Several important characteristics result from the fact that digital multitesters are electronic. One characteristic is that they are basically voltmeters, and when used as ammeters or ohmmeters, the circuit arrangement is such that the DMM is used as a voltmeter. A second characteristic is that the DMM has a high internal resistance which is present on all functional ranges. As will be shown in Chapter 3, having a high internal resistance is a very desirable characteristic because circuit loading is usually negligible when using a DMM.

The third and fourth characteristics are due to the digital display. Because the display is digital, there is neither parallax error nor interpolation error as may occur when reading the scale on an analog meter. Also, because the display is digital, the conversion accuracy is within ±one digit on any of the scales used. The accuracy due to the display remains constant over all ranges and does not vary. As a result, overall accuracies of DMMs are typically 0.05% to 1.5% as compared to 3% to 4% for analog meters.

The fifth characteristic relates to the special functions that are available. An audio tone or a buzzer that sounds when measuring circuit continuity is a common example. A special check for semiconductor junctions is available on the meter in *Figure 2-1*, as well as the "data hold" feature discussed previously. Other meters may have several other special functions such as an audible overrange, conductance measurement, temperature and capacitance measurements and, along with the semiconductor diode junction test, some may also measure the gain or beta (h_{FE}) of a transistor.

MAKING SIMPLE MEASUREMENTS WITH THE DMM

Measuring a DC Voltage

A very common device around most homes is a small power adapter that converts ac line voltage (115VAC) to a dc voltage. They are usually used on small radios, tape recorders, wireless telephones, calculators or for recharging small power tools. Using a DMM for measuring the output voltage of a typical power adapter is shown in *Figure 2-3.* The ON-OFF switch is set to the ON position and the FUNCTION/RANGE switch to the DC V function and the 20V range. The test leads are plugged into the appropriate jacks — black to COM and red to V-Ω-mA. The probe tips of the test leads are then connected to the power adapter plug as shown in the figure. Make sure that the adapter is plugged into an ac outlet. The voltage polarity of the adapter output is determined by the particular piece of equipment using the adapter; however, when measuring with a DMM, you do not have to observe polarity. The DMM will automatically display the polarity and the value of the voltage of the adapter output. If the adapter is not plugged into its equipment, it is said to be unloaded, and the voltage delivered at its output plug may be more than its rated

Figure 2-3. Measuring a DC Voltage with a DMM

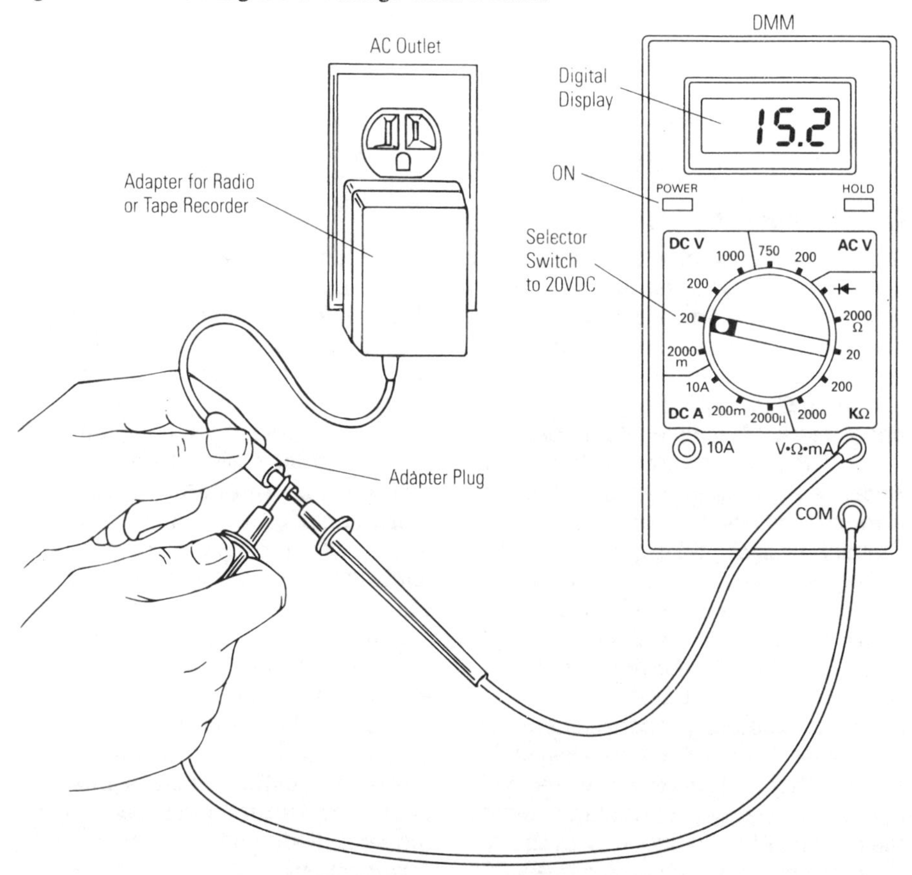

voltage; that is, the value printed on the adapter. For example, an adapter with a rated voltage of 9V may have an output of as much as 15V when it is unloaded. Most are full-wave rectifiers; however, some adapters have half-wave rectifiers, which would result in a lower rms voltage reading. If you can open the equipment and measure the voltage at the adapter connection inside the equipment, then the voltage reading under load should be close to the rated value. If you read 0 volts or a very low value, the adapter is probably defective. When you get a replacement, be certain to get one that has the same output voltage and current rating and the same type of output connector.

Measuring an AC Voltage

In Chapter 1 you saw how an analog VOM is used to measure the ac voltage at an ac outlet. Since this is probably the most common source of electrical power around the house, let's look at the procedure again, but this time using a DMM.

WARNING

The ac line voltage from a wall outlet is very dangerous. Always hold the test probe far back from the metal tip. *Never touch the metal part of the probes when measuring voltage with the multitester.* Wear rubber-soled shoes, if possible, and do not stand in water or on damp ground or floor when making these measurements. If possible, use only one hand to make the measurement. The old rule, "keep one hand in your pocket," is still a good one.

Measuring the ac voltage at a household outlet is shown in *Figure 2-4.* The ON/OFF switch is set to ON and the FUNCTION/RANGE switch is set to the AC V function and to the full-scale range of 200V. Carefully insert the test probe tips into one contact on each of the plug outlets as shown in the figure. ***Do not let the metal probe tips touch together.*** The power line voltage fluctuates around a nominal 115VAC, so the reading on the digital display should be between 110V and 125V.

Now move one probe to the screw that fastens the outlet face plate in place, and measure the voltage from each outlet contact to the screw *(Figure 2-4.)* Make sure the screw is metal and that you make good contact. The outlet is grounded if a voltage is measured between either outlet contact and the screw. Chapter 7 discusses the importance of grounded outlets.

Measuring a DC Current

A simple use of the dc current measuring function is to measure the "current draw" of several devices on an automobile. A current meter is connected *in series* with the element or device drawing the current, *never across* the device or the power source. To avoid blowing the DMM's 2A fuse, start by using its 10A range until you are sure that the current draw is less than 200mA (0.2A). This requires that the red test lead be in the 10A jack and the FUNCTION/RANGE switch be set to the DC A function and to the 10A full-scale range. If the DMM reads less than 0.2A on the 10A range, go to a lower range to make the final reading. *Do not crank the engine or operate any accessory that draws more than 10A. You could possibly damage the meter beyond repair because the 10A range is not fused and is rated to handle an overload of 12A for only 60 seconds.*

Figure 2-4. Measuring an AC Voltage (115VAC Outlet) with a DMM

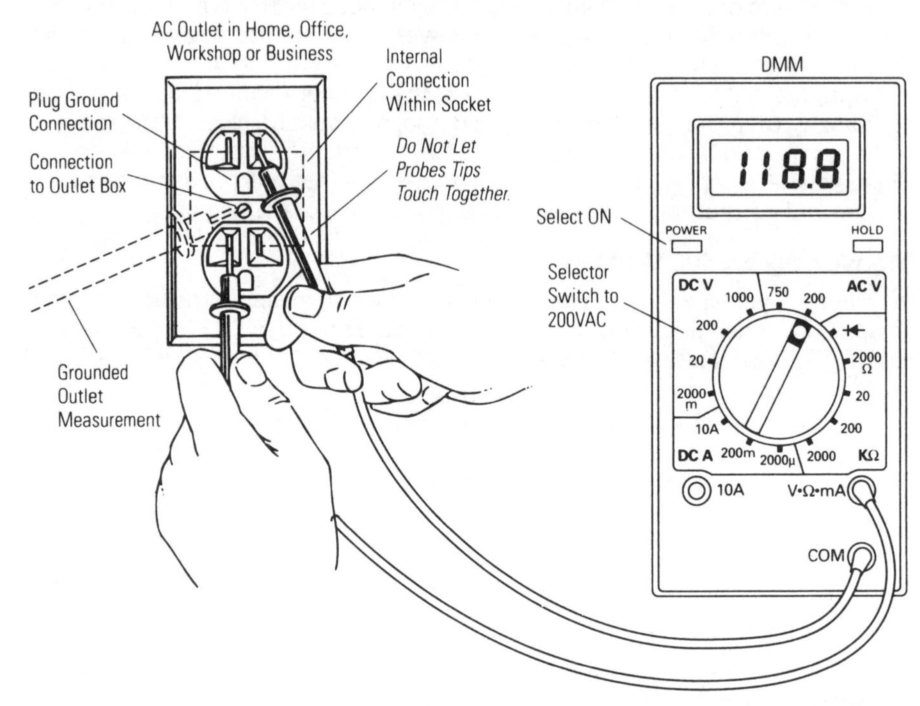

A measurement to check the current draw from the battery of the entire automotive electrical system is shown in *Figure 2-5.* Begin with the ignition switch off and all the automobile accessories off. Also, be sure the headlights and parking lights are off. Connect the DMM in series with the battery at the cable connector. Disconnect the battery cable connector from the positive battery post and connect the DMM as is shown in *Figure 2-5.* Always begin with the 10A range and decrease to an appropriate range after you turn on an accessory and determine that it is drawing less current than the full-scale value of the next lower range.

WARNING

Do not turn the ignition switch to the START position. Your DMM probably will be severely damaged if you do.

Turn the ignition switch to the position where power is supplied only to accessories. Depending on the complexity of your car's electrical system, there may be several electrical systems that are automatically turned on and off depending on whether the door is open or closed, the position of the ignition switch, and whether the seat belts are fastened. The engine computer system and other digital circuits, including the clock and the memory in the radio, may always draw a certain small amount of current from the battery. Also, remember that the dome light, the trunk light and the underhood light may be on while you are working on the car. A piece of tape over the switch button on the door post will turn off the dome light and the ignition-key-on warning bell. The trunk light and underhood light are usually controlled by a "tilt" switch, but

Figure 2-5. Measuring DC Current Drawn from an Automobile Battery

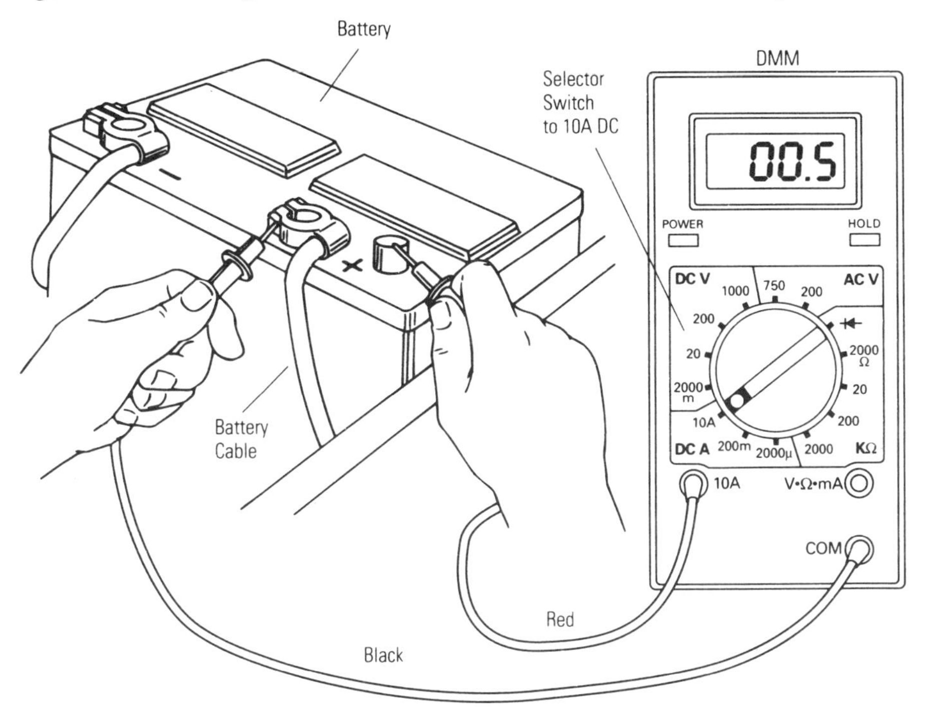

you can remove the lamp to keep its current from being included in the DMM's reading. The seat belt warning bell should automatically turn off after a few seconds. Write down the initial current reading on the DMM for later use.

Turn on one accessory, such as the radio, air conditioner, heater, rear-window heater, cruise control or headlights. (Some of these accessories may operate only with the ignition switch in the ON position, so they may not operate with the ignition switch in the ACC position.) Adjust the full-scale range to the lowest full-scale range available for the measured value. Read the current directly, write it down, and turn off the accessory. From this reading, subtract the initial current reading you noted earlier to obtain the actual current drawn by the accessory. Repeat this procedure for each accessory to obtain the individual accessory currents. If you want to know the total amount drawn, just sum the individual current values. (You could leave each accessory on and measure the total current when all of them are on, but be careful you don't exceed the maximum current rating of the DMM. This method will not be as accurate because the battery voltage will drop slightly as more accessories are turned on.)

Measuring an AC Current

Rarely is an analog VOM capable of measuring ac current directly, and not all DMMs have the function either. For example, the DMM shown in *Figure 2-1* measures dc current, but in order to measure ac current, an ac voltage is measured across a small resistor placed in the circuit.

Figure 2-6 is an example of measuring an ac current with a DMM in the low-voltage part of a door chime circuit. If the system is working properly, the transformer secondary current and secondary voltage can be measured at the door chime push-button switch. By removing the two screws that hold the switch to the door facing, you can access the two connections on the back of the switch.

First, use the DMM to measure the secondary voltage. Set the ON-OFF switch to ON. Set the FUNCTION/RANGE switch to the AC V function and to the full-scale range of 200 V. Plug the test leads into the COM and V-Ω-mA jacks to make a voltage measurement. Connect the test lead probes across the push-button switch terminals. Since the switch is open (not pushed), the DMM should read between 10V and 25V, depending on the transformer secondary voltage.

To measure the current that the door chime draws, the DMM must be set up to measure an ac voltage. Set the ON/OFF switch to ON. Set the FUNCTION/RANGE switch to the AC V function and to the full-scale range of 200V. Connect a 10-ohm resistor across the push-button screw terminals as shown in *Figure 2-6c*. Plug the test leads into the COM and V-Ω-mA jacks. The door chime circuit is completed through the resistor. The ac voltage measured across the resistor is divided by 10 ohms to determine the current. The current should be about 0.5A. There is no need to press the push button; in fact, if you do, you will short out the resistor and the voltage reading on the DMM.

Measuring Resistance

Have you ever wondered if a light bulb was good or not? You can find out without trying it in a lamp base by using your DMM or VOM ohmmeter to check its filament resistance. (This method will not work on fluorescent lamps.) *Table 2-1* gives typical values of resistance for the filaments of several common wattage, 120VAC light bulbs. You can make a similar table for light bulbs that you commonly find around your house or work place by measuring a known good one and writing down the resistance. In fact, a table of the "good" resistances of many items can be very useful if you later have a need to check one of them.

Table 2-1. Resistance of Typical 120VAC Lamp Filaments

Wattage of Bulb (W)	25	40	60	75	100
Resistance of Filament (Ohms)*	46	26	17	13	10

* This is the resistance of the cold filament. The value calculated by Ohm's law when the filament is hot will be slightly higher.

Figure 2-7 shows how to measure the filament resistance of a typical 120VAC light bulb (note that this measurement is of a cold filament). Making a resistance measurement is somewhat different than making a voltage measurement or a current measurement.

CAUTION

Never attempt to measure resistance in a live circuit. Always disconnect the part to be measured from the source of power before you attempt to measure its resistance.

Figure 2-6. Measuring AC Current and Voltage to Door Chime

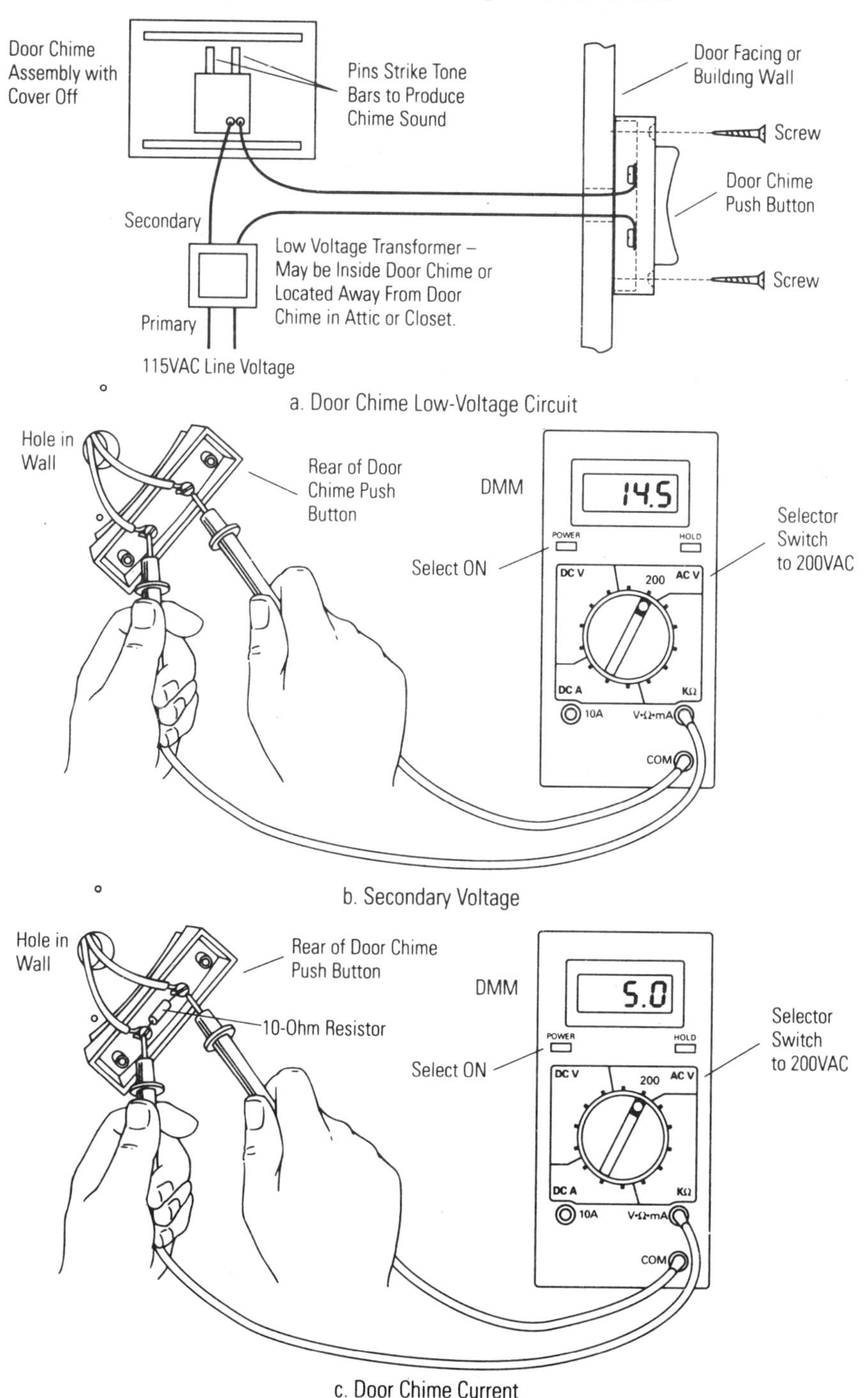

Figure 2-7. Measuring a Light Bulb's Filament Resistance

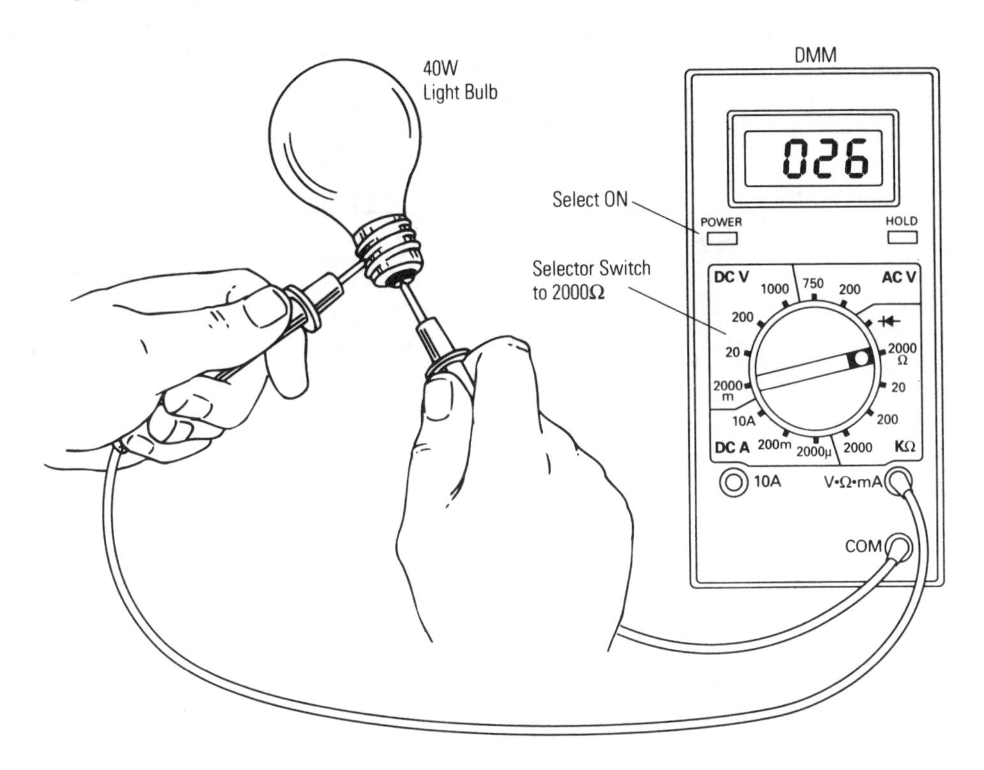

Set the ON-OFF switch to ON. Set the FUNCTION/RANGE switch to the Ω (ohms) function and the full-scale range of 2000Ω. Plug the test leads into the COM and V-Ω-mA jacks. Touch the test lead probes to the lamp connections and read the resistance directly in ohms. Of course, if the light bulb is burned out, there will be an open circuit and a 1 will be displayed.

Continuity Tests

To measure continuity (complete circuit) or alternatively, no continuity (open circuit), the DMM is used as an ohmmeter. The FUNCTION/RANGE switch is set on the lowest resistance range.

One of the main uses of the continuity function is to answer the question: Is there a complete circuit? Or, the other alternative: Is there continuity when there should not be; that is, a short? *Figure 2-8* shows a handy use for the continuity check — testing extension cords or appliance power cords to make sure that they are providing good connections. Set the ON-OFF switch to ON. Set the FUNCTION/RANGE switch to the Ω (ohms) function and to the 2,000Ω position.

There are two tests to make. (1) Test the continuity of both (or all three) of the wires in the power cord from end-to-end. (2) Then, as shown dotted, test to see that the wires do not have continuity to each other; that is, that they are not shorted.

Figure 2-8. Testing a Power Cord to Make Sure Connections Are Good (Good Continuity)

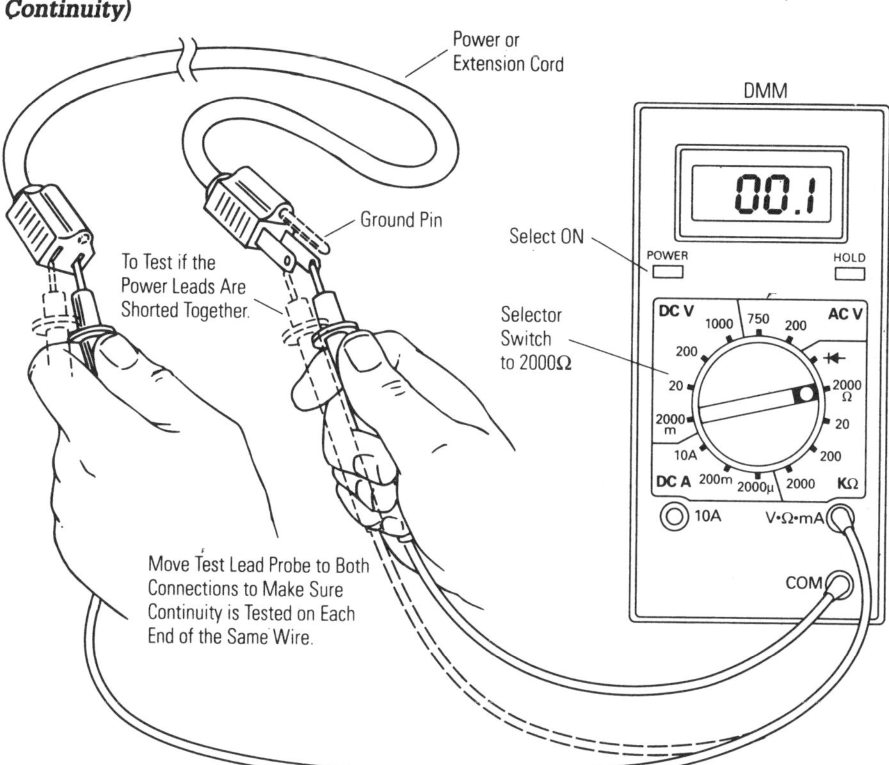

Often DMMs have an audible continuity function. With the DMM set to measure resistance and the CONTINUITY function activated, when the test lead probes are touched together, the DMM emits an audible sound. This electronic sound can be used to look for short circuits or for tracing for an open circuit. Any time the continuity circuit resistance is less than a minimum amount, usually 20Ω to 300Ω, the DMM emits an audible sound when the circuit is complete.

Measuring Semiconductors

Sometimes you may want to know if single diodes or transistors out of a circuit are good. There are two basic tests using a DMM that can be made on semiconductors to provide the answer. (More advanced tests are described in Chapter 4 and Chapter 5 shows how to test semiconductors that are connected in a circuit.) The two basic tests are made on single diodes and single bipolar junction transistors. Let's first consider the diode test.

Diode

The diode should be tested both with forward voltage and reverse voltage applied. *Figure 2-9* shows an example. The ON-OFF switch is set to ON. The FUNCTION/RANGE switch is set to the Ω (ohms) function and to the ⏴

(diode test) position. When the test leads are connected with reverse voltage; that is, red to cathode and black to anode, the display should show a 1. Any other display indicates a defective diode.

Figure 2-9a shows a good diode being tested in the forward voltage direction with the DMM displaying the forward bias voltage of the particular diode. This is a very interesting feature of the DMM because you can tell something important about a diode by knowing its forward bias voltage. For example, a germanium diode will have a forward bias voltage of about 0.2V, a silicon diode about 0.6V, and a gallium arsenide (light-emitting) diode about 1.4V. *Figure 2-9b* shows a good diode being tested in the reverse voltage direction with the DMM displaying a 1 which indicates the diode is good.

Figure 2-9. Measuring a Semiconductor Diode

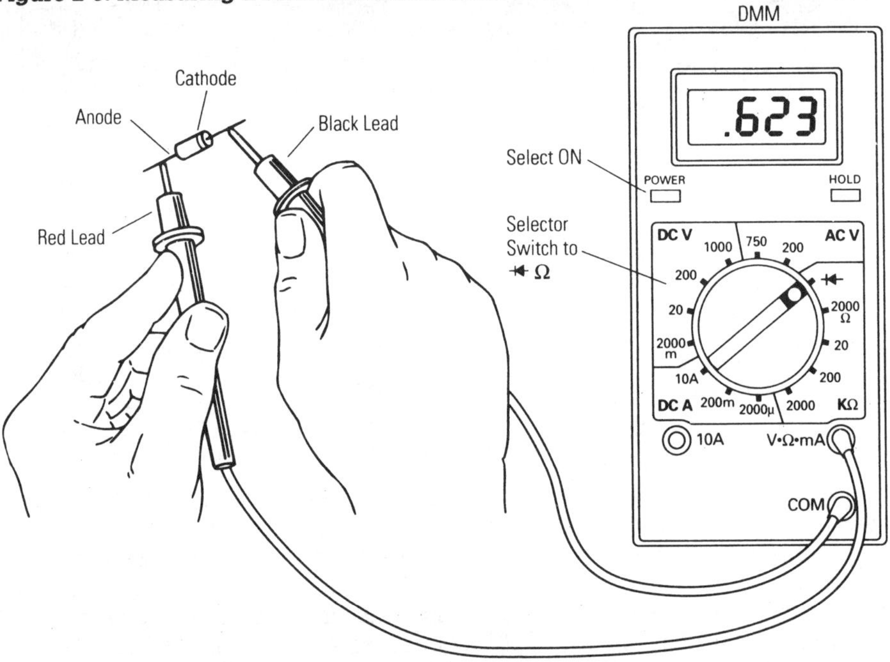

a. Forward Voltage Condition

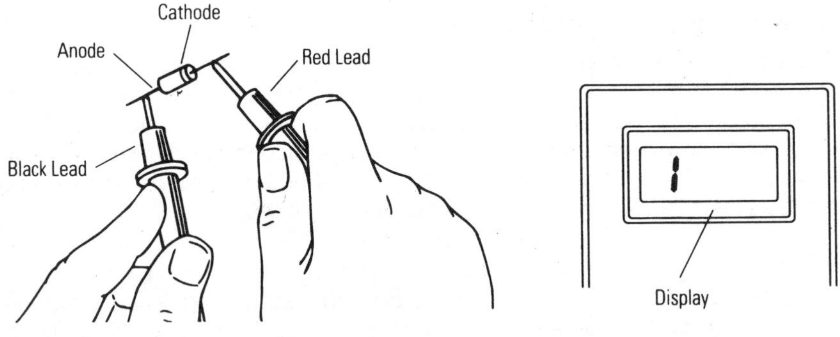

b. Reverse Voltage Condition

The DMM shown in *Figure 2-1* will display a forward voltage value of up to about 2.8V with a forward current of up to 3mA. If in the forward voltage test the display shows 1, then the diode is open; if in the forward voltage test the display shows 000, then the diode is shorted. In either case, the diode is defective. Now, let's consider the transistor.

Transistor

A transistor's junctions can be tested like two diodes, but because of the different types of transistors (NPN or PNP) and the multiple junctions, the explanation is a bit complicated to discuss at this time. (We will explain it in Chapter 4.)

HOW A DMM WORKS

A block diagram of a typical digital multimeter is shown in *Figure 2-10.* Either resistance, dc voltages or ac voltages can be measured. *Figure 2-10* shows a DMM measuring a dc voltage.

The test probes make the connections to the dc voltage, which is brought into the DMM through a signal conditioner, and coupled into a circuit called an analog-to-digital converter (ADC). The ADC accepts a voltage and changes it to a digital code that represents the magnitude of the voltage. The digital code is used to generate the numerical digits that show the measured value in the digital display.

Figure 2-10. Block Diagram of a Basic Digital Voltmeter

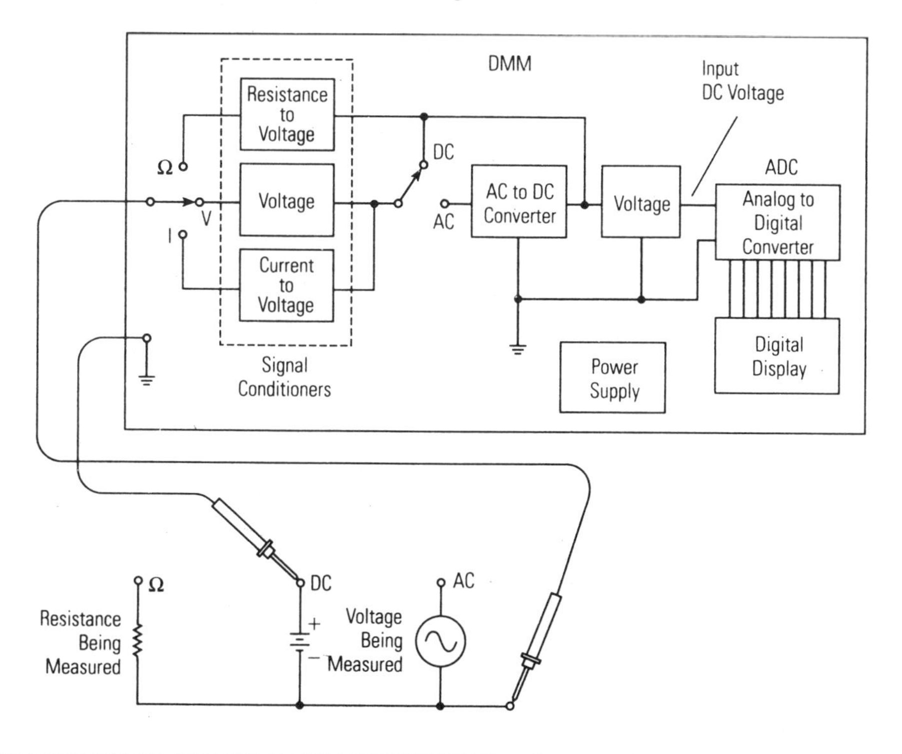

Digital Display

An easy way to understand how a DMM works is to begin at the display. *Figure 2-11a* shows a common way to display a numerical digit. A 7-segment array of elements is used to form the digits. The array elements may be vacuum fluorescent, electroluminescent, or plasma display elements; however, the displays for portable DMMs used in the laboratory or service shop are often light-emitting diodes (LEDs) or liquid crystal displays (LCDs). Liquid crystal displays use very little power, but must have ambient light or back illumination to operate the display. LED displays use much more power, but are much brighter and not easily washed out in bright sunlight.

Figure 2-11. 7-Segment Digit Display

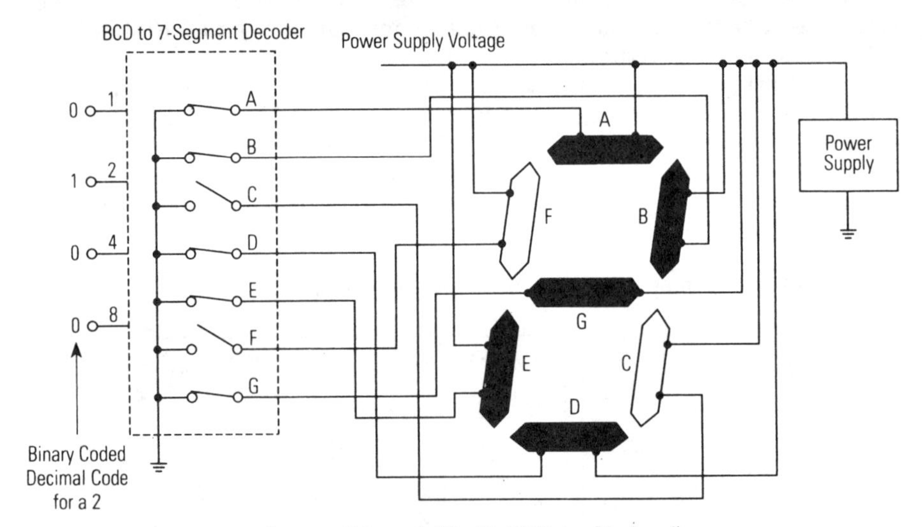

a. Segment Schematic (The Digit 2 Being Displayed)

Digit	Segment							BCD Code
	A	B	C	D	E	F	G	8421
0	●	●	●	●	●	●	●	0000
1		●	●					0001
2	●	●		●	●		●	0010
3	●	●	●	●			●	0011
4		●	●			●	●	0100
5	●		●	●		●	●	0101
6			●	●	●	●	●	0110
7	●	●	●					0111
8	●	●	●	●	●	●	●	1000
9	●	●	●			●	●	1001

b. Excitation Table for Digits

As shown in *Figure 2-11a,* (which is a typical LED display schematic), a power source is connected to each LED segment, and each LED segment is excited by passing current through it. In the example shown, the segments necessary to display the numeral 2 are grounded and current through them causes the segments to emit light. The table shown in *Figure 2-11b* lists the segments that must be excited to display any of the numerals from 0 to 9. As shown in the example of *Figure 2-11a,* a decoder establishes the proper segments that must be grounded. The decoder has an input digital code that represents the numeral required. As shown in *Figure 2-10,* the binary-coded-decimal code comes to the digital display from the ADC.

SIGNAL CONDITIONERS

Because the ADC always needs an input dc voltage, all uses of the DMM as a voltmeter, ammeter, or ohmmeter requires that signal conditioners be used to convert the measured quantity into a dc voltage. The three types of conditioners are shown in *Figure 2-10.*

BASIC CHARACTERISTICS OF DMMs

Specifications define the characteristics of DMMs. The most common are accuracy, internal resistance (input impedance), range, display, resolution, response time, and protection. The most unique specifications are the ones that define the digital display and the full-scale range.

Digits of the DMM Display

A strange term was coined to specify a DMM's digital display. The term "half-digit" is used to describe the display capability of the DMM and the reading beyond full-scale that it can display. This is called overranging.

If a DMM is classified as a 3-1/2-digit DMM, it means that the full-scale reading is displayed in three digits and that the digit to the left of the 3 digits for full-scale is restricted in range. For example, 0.999 would be the full-scale reading when the DMM is measuring 1V on the 1V range. The 1/2-digit specification means that the full-scale digit is restricted to a 1. A 4-1/2-digit DMM has the capability to read a value to 19999. A 3-3/4-digit DMM has the capability to read a value as high as 3999. Therefore, the 1/2 or 3/4 digit specifies the overranging that the DMM can read when set on a particular range. The 1/2 digit is by far more common than the 3/4 digit overrange. The overrange digit is sometimes referred to as a "partial" digit.

Full-Scale Range

Range is specified in one of two ways: (1) a full-scale range with usable overrange capabilities specified as a percentage, typically 100%; or (2) full scale specified as the maximum possible reading encompassing all usable ranges, often 1.999. For example, a DMM may be specified as having 1V full scale with 100% overrange, thus indicating useful operations to 2V; or the same DMM may be simply specified as having a 2V scale.

Some ranges may not be used to full scale. For example, the higher voltage ranges may be limited because of the voltage breakdown of internal components. A DMM with a 1999V full-scale range may not be used to read over 1000V because the divider resistors cannot withstand over 1000V without breakdown. The capability may be even lower on ac because of peak voltage.

Overranging

Overranging was instituted to take full advantage of the upper limits of a range. It also provides better accuracy at the top of a range (see the example in the next section). In other words, overranging allows a DMM to measure voltage values above the normal range switch transfer points without having to change ranges. It allows the meter to keep the same resolution for values near the transfer points.

An analog VOM is easily damaged by an overload or overrange condition which drives its indicating needle off-scale and against the pin; however, the DMM usually has a method of using the overrange feature by a certain amount and then indicating an overload above this overrange condition without damage to the instrument. A common method of indication of overload is to cause the most significant digit to blink and to blank the three least significant digits of the display. Some DMMs have more elaborate overrange condition indicators. For the DMM in *Figure 2-1*, an input overrange condition sets the most significant digit to a 1 and blanks the three least significant digits. The polarity is usually displayed in front of the overrange symbol. On some DMMs a buzzer sounds when either voltage or current measuring ranges are overloaded. Overrange is not used on resistance ranges.

The extent to which the overrange is possible is expressed in terms of the percentage of the full-scale range. For example, a 3-1/2-digit DMM with 20% overrange can display voltages up to 1.199V on its 1V range. Overranging from 5-300% is available depending upon the make and model of the DMM.

Accuracy and Resolution

Simple accuracy specifications are given as "plus/minus percentage of full scale, plus/minus one digit." The "plus/minus one digit" portion of the specification is caused by an error in the digital counting circuit. The "plus/minus percentage of full scale" includes ranging and ADC errors.

The "plus/minus one digit" also relates to the resolution of the DMM. The resolution of an instrument is directly limited by the number of digits in the display. A 3-1/2-digit DMM has a resolution of one part in 2000 or 0.05%, which means that it can resolve a measurement of 1999mV down to 1mV. A 4-1/2-digit instrument has a resolution of one part in 20,000, or 0.005%.

A 2-1/2-digit or a 3-digit DMM is considered significantly better than a good VOM multitester as far as accuracy and resolution are concerned. Accuracy generally lies between 0.5% and 1.5% and resolution to 0.5%. The 3-1/2-digit and 4-1/2-digit meters generally have accuracy one order of magnitude higher; that is, between 0.5% and 0.05% with 0.05% resolution. A resolution and accuracy of this amount will generally suffice for most service work today. A 4-1/2-digit and 5-1/2-digit DMM generally indicate an accuracy of 0.05% and better with a resolution of 0.005%. These are indeed considered laboratory instruments. They usually are specified with a "plus/minus percent of reading, plus/minus percent of full scale, plus/minus one digit" specification. They also may have specifications that qualify the accuracy at temperatures other than 25°C.

An example will show how overranging and the number of digits of resolution work together. Suppose that a voltage that is being measured changes from 9.999V to 10.026V. A 4-digit DMM without overranging could measure the original voltage as 9.999V; however, to measure the new value would require a change of range which would result in a reading of 10.02V. The additional change of 0.006V would not be displayed. With overranging, however, the second voltage would be displayed as 10.026V and there would be no loss of resolution.

A DMM has essentially the same accuracy on ac that it does on dc voltage measurements while the accuracy of an analog meter is most assuredly less accurate on the ac voltage measurements. Accuracy also depends upon the frequency response or bandwidth of the DMM, and on the ac waveshape when the meter does not measure true RMS voltages.

Input Impedance

DMMs have an input impedance of at least 1MΩ, and more commonly, 10MΩ. This holds true on dc measurements and on ac measurements over the frequency range specified for the DMM. The high input impedance helps to maintain the higher accuracy of measurement provided by the DMM because the measured circuit loading (which we will cover in Chapter 3) is considerably less.

Response Time

Response time is the number of seconds required for the instrument to settle to its rated accuracy. The response time consists of two factors: (1) the basic cycle rate of the DMM's internal analog-to-digital converter; and (2) the time required to charge capacitances in the input circuit. Instead of response time, some manufacturers simply give a number of conversions per second.

Protection

Meter protection circuits prevent accidental damage to the DMM. The protection circuit allows the instrument to absorb a reasonable amount of abuse without affecting its performance. The specification of input protection indicates the amount of voltage overload which may be applied to any function or range without damage. AC RMS overload limits are usually smaller because of the peak voltages that occur. A separate dc limit may be indicated to protect the input coupling capacitor used for ac measurements from breaking down. Overloads from sources outside the specified frequency range of the instrument may not have as great a protection range. The current measuring circuitry is usually protected by a fast-blowing fuse in series with the input lead.

More on Displays

We have already mentioned LED and LCD displays. The light-emitting diode (LED) display has been popular in DMMs because of its brightness, excellent contrast and low cost; however, its high power dissipation is a disadvantage for battery-operated units. The liquid crystal display (LCD) is becoming very popular because of its very low power drain and decreasing cost. The newer types do not wash out in bright sunlight. However, a disadvantage is that many of them will freeze at fairly moderate temperatures and become completely useless. The units must be kept above 32°F (0°C).

SPECIAL FEATURES

Auto Polarity

The automatic polarity feature further reduces measurement error and possible instrument damage because of overload due to a voltage of reverse polarity. A + or – activated on the digital display indicates the polarity and eliminates the need for a polarity switch setting or for reversing the leads.

Hold That Reading!

The digital multitester shown in *Figure 2-1* has a hold feature that is operated remotely by means of a special HOLD button included on the front panel. It is particularly useful when making measurements in a difficult or hazardous area because the operator can keep his eyes on the test lead probes while the reading is captured and held. Then he can remove the probes and read the value.

In some DMMs, the hold signal sets an internal latch that captures the data. When the DMM gets to the point in its cycle where data is to be displayed, the state of the internal latch is sampled. New data is not transferred while the latch is in the hold state. In other DMMs, the hold feature stops the instrument clock. The last value displayed on the LCD remains displayed until the ground is removed from the input terminal.

SUMMARY

Since we now know something about analog and digital multitesters and how to use them, let's look further into the application of multitesters to make dc and ac measurements.

CHAPTER 3

DC and AC Measurements

This chapter deals with how to make measurements with a multitester, whether it is with an analog meter as described in Chapter 1, or with a digital meter as described in Chapter 2. However, before discussing the actual measurements, let's review what electricity is and the three basic fundamentals of dc circuits: Ohm's law, series and parallel circuits, and polarity.

WHAT IS ELECTRICITY?

Matter is thought to consist of extremely tiny particles grouped together to form atoms. In the study of electricity, the two particles of interest are the positively charged proton and the much smaller negatively charged electron. They exert forces on each other and, basically, the forces cause the charges (mostly the electron) to move. So *electricity is a form of energy produced by charges moving under the influence of an electromotive force*, and *charge moving through an electric circuit is an electric current*. Charge that is not moving (stationary) is static electricity.

OHM'S LAW

In the early 1800s, a German physicist, Georg Simon Ohm, discovered the basic relationship of voltage, current and resistance in an electrical circuit which he expressed as:

$$I = \frac{V}{R}$$

Stated in words, it says that the current I in a dc circuit varies directly with the voltage, V, applied to the circuit and inversely with the resistance, R, in the circuit. With a given resistance, if voltage is increased, current will increase; with a given voltage, if resistance is increased, current will decrease.

SERIES CIRCUITS

In a series circuit like the one shown in *Figure 3-1,* there is only one current path. As shown, the *conventional current* direction is through the circuit from the positive terminal of the battery to the negative terminal. *Electron current* is in the opposite direction — the negatively charged electrons are attracted to the positive terminal of the battery and released from the negative terminal.

Figure 3-1. Simple DC Series Circuit

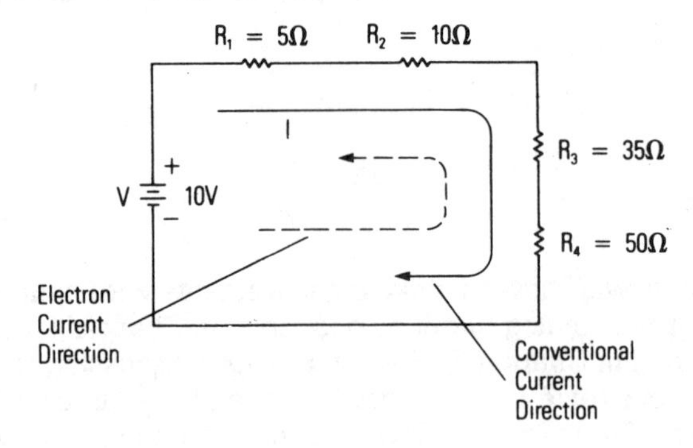

The amount of current in the series circuit can be found by using Ohm's law. First, find the total resistance by adding the values of all the resistors in the circuit — the sum is 100 ohms. Next, divide the voltage (10 volts) by the total resistance (100 ohms):

$$I = \frac{10V}{100\Omega}$$

$$I = 0.1A$$

The current is 0.1 ampere. There is the same current throughout the series circuit and through each series circuit component. The current has only one path through all of the circuit components in series. This is a basic characteristic of a series circuit: **the current is the same through all components in a series circuit.**

PARALLEL CIRCUITS

Parallel circuits are different. Current in a parallel circuit has more than one path. The different paths are called branches. *Figure 3-2* is a simple parallel circuit. As with the series circuit, it has a 10-volt battery to supply the current, but the current is through two branches — one through R1 and the other through R2. The total current I_T from the battery divides into the two branch currents, I_1 and I_2. **The total current I_T is the sum of the two branch currents.** Therefore:

$$I_T = I_1 + I_2$$

Each branch current can be calculated by using Ohm's law:

$$I_1 = \frac{10V}{100\Omega} = 0.1A$$

and:

$$I_2 = \frac{10V}{400\Omega} = 0.025A$$

Figure 3-2. Simple Parallel Circuit

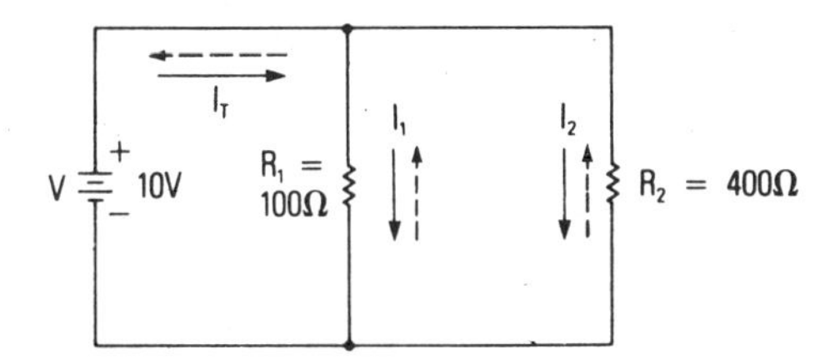

The total current is:

$$I_T = 0.1A + 0.025A$$
$$I_T = 0.125A$$

Using powers of ten, the total current of 0.125A can be expressed as 125×10^{-3} ampere. Since the prefix "milli" is equivalent to 10^{-3}, this current can be expressed as 125 milliamperes. The common abbreviation for milliamperes is mA, thus, the total current can also be expressed as 125mA. So the current through the R_1 branch is 100mA, the current through the R_2 branch is 25mA, and the total current is 125mA.

Note that the voltage across each branch is the same. This is a basic characteristic of any parallel circuit: **the voltage is the same across each branch of a parallel circuit**.

POLARITY

When making dc measurements with your multimeter or multitester, particularly an analog meter, you must be concerned about observing polarity when you connect the test leads to the circuit. If correct polarity is observed, the test leads will be connected properly, and current through the meter will be in the direction to cause an analog dc meter's needle to deflect up scale (almost always to the right). With incorrect polarity, the meter needle will deflect down scale (almost always to the left) and "peg" against the stop. This can cause significant damage to the meter. Digital voltmeters do not have this problem. They simply display the polarity of the voltage as part of the digital readout. With both types of meters, the reference point for the measurement is very important.

Reference Point

A reference point is a point in a circuit to which the voltage at other points in the circuit is compared. The phrase "with respect to" is commonly used to refer to the chosen reference point. The concept of a reference point is demonstrated best by an example.

Figure 3-3 combines a pictorial and schematic of the common electrical system for starting a car. The negative terminal of the battery is connected to the automobile chassis, which becomes a connecting link of the circuit. It is like a common wire connecting all points marked "chassis ground." It becomes an excellent reference point for all measurements in this automotive circuit.

Figure 3-3. Automotive Starter Circuit

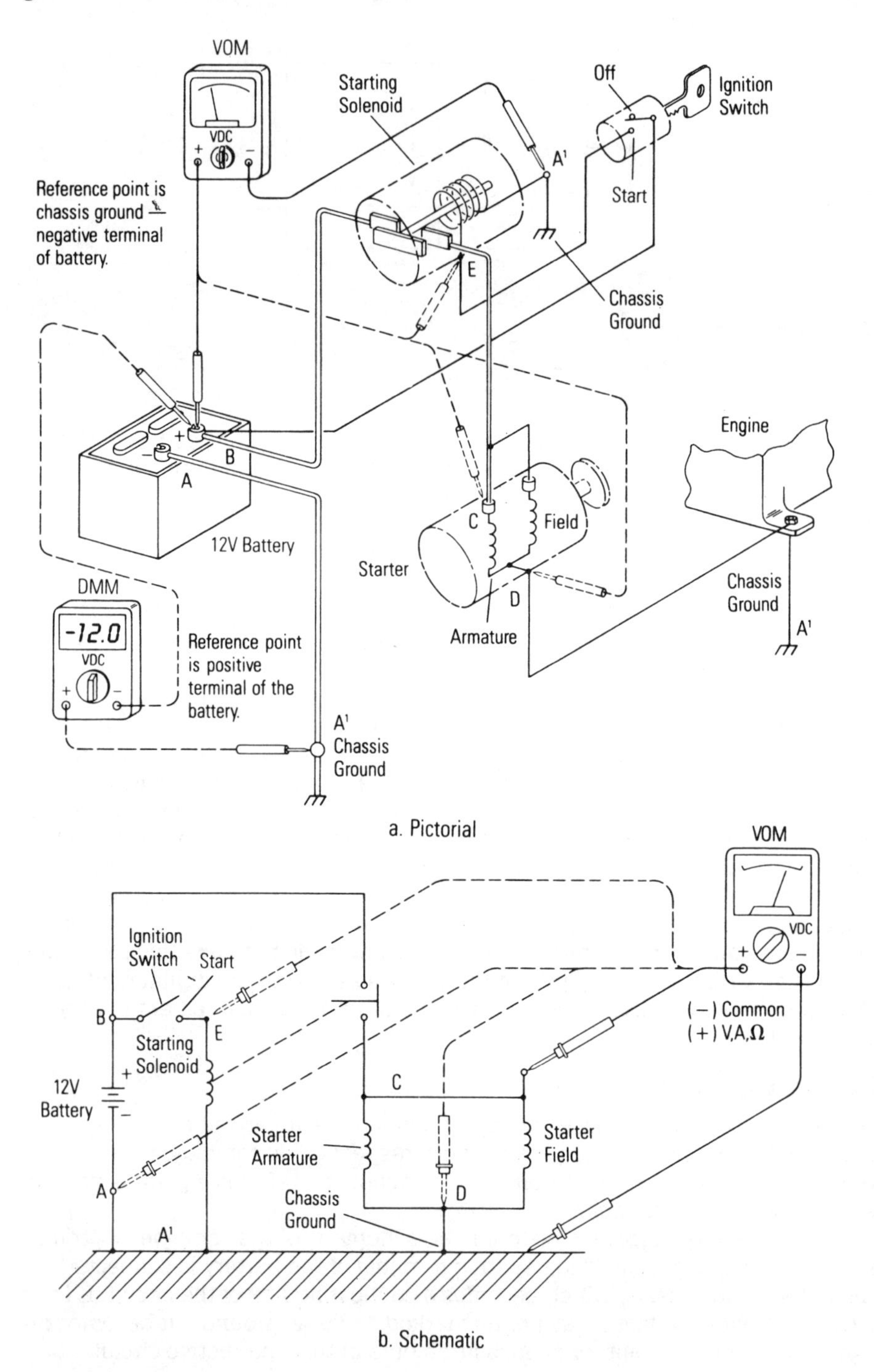

As shown in *Figure 3-3*, the common or negative terminal of a VOM is connected to the negative terminal (A) of the battery through chassis ground (A'). A voltage measurement is made at point B to measure the battery voltage by placing the positive probe of the meter on point B. The meter needle deflects up scale and indicates +12 volts, thus, point B is 12 volts positive *with respect to* the reference point, which is chassis ground and the negative terminal of the battery. The following chart gives the other voltages in the circuit *with respect to* the reference point. For all measurements, the common (−) probe of the meter is on A', chassis ground (or the negative battery terminal).

Point	Meter Reading	Comment
A	0V	Both probes on chassis ground
C	0V	When ignition switch is OFF
C	+12V	When ignition switch is on START
D	0V	Both probes on chassis ground
E	0V	When ignition switch is OFF
E	+12V	When ignition switch is on START

Note that when a voltage measurement is made, the meter is across two points. In all cases above, all points are a positive voltage *with respect to* the reference point, or else they have the same potential as the reference point (when the voltage read is 0 volts).

Other reference points could have been chosen. *Figure 3-3* also shows the case when the common probe of a DMM is connected to the + terminal (B) of the battery. Now the measured voltages at the various points are as shown in the following chart compared to the reference point B.

Point	Meter Reading	Comment
A	−12V	Reference point now B
B	0V	Both probes at same point
C	−12V	When ignition switch is OFF
C	0V	When ignition switch is on START
D	−12V	Reference point now B
E	−12V	When ignition switch is OFF
E	0V	When ignition switch is on START

Note that the negative terminal of the battery (A) and point D, which are connected to chassis ground, now have a negative polarity (−12V) *with respect to* the reference point, which is the positive terminal of the battery (B). Both A and D were at 0 volts when the reference point was the negative terminal of the battery (A). **Remember: The polarity of a point in the circuit depends on the point chosen for the reference point.**

Polarity of Voltage Drops

One other main point on polarity relates to the polarity of voltage drops across components in a circuit. Refer to *Figure 3-4* where three resistors and a battery are in a series circuit. The sum of the voltage drops across the resistors equals the battery voltage. The end of any resistor which is closer to the positive terminal of the battery is positive with respect to the negative terminal (the reference point). Conversely, the end of any resistor closer to the negative terminal of the battery is negative with respect to the positive terminal (the reference point). However, if a reference point between the negative and positive terminals is chosen, the polarity of the voltage drop depends on the reference point.

For example, if a DMM is connected with its common probe at the intersection of R_2 and R_3 (D) as shown in *Figure 3-4,* then the voltage drop across R_2 at A is positive with respect to D, and the voltage drop across $R_1 + R_2$ at B also is positive with respect to D. However, the voltage drop across R_3 at C is negative with respect to D.

If the common probe of the DMM is moved to point A (not illustrated), it becomes the reference point, and the polarity of voltage drops with respect to A are as given in the following chart.

Point	Voltage Drop
B	V_{R1} is positive
D	V_{R2} is negative
C	$V_{R2} + V_{R3}$ is negative

With this review of Ohm's law, series and parallel circuits, and polarity, let's move on to look at multitester measurements.

Figure 3-4. Polarity of Voltage Drops

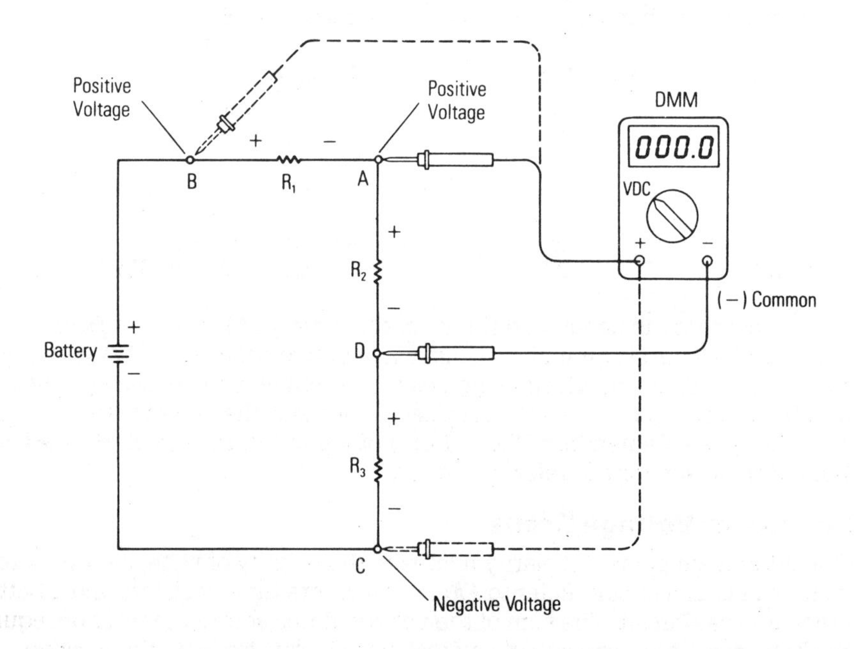

METER LOADING

Meter loading can cause inaccurate voltage measurements. The concept of meter loading is explained best by looking at the example illustrated in *Figure 3-5.* In *Figure 3-5a,* a VOM is used to measure the voltage across R_2 in a simple dc series circuit which is supplied current by the voltage V. Notice that to make the voltage measurement you must place the voltmeter across resistor R_2. The common (−) probe of the meter is at A and the (+) probe of the meter is at B. The VOM is represented by a resistor in series with a meter movement. The value of the resistance is equal to the input impedance of the meter.

Meter Forms Parallel Branch

As shown in the equivalent circuit in *Figure 3-5b,* placing the VOM across R_2 forms a parallel branch of the circuit with R_2. It adds another path for current through R_M besides the path through R_2. It means that I_T increases because now I_T equals I_1 plus I_2 rather than just I_1 without the meter present. How large is I_2? As discussed in Chapter 1, if the voltage drop across R_2 causes a full-scale reading on the chosen meter range, the current I_2 through R_M will be equal to the current required to deflect the meter movement to full scale. R_M in parallel with R_2 reduces the equivalent circuit resistance represented by the parallel branch, and produces a corresponding error in the voltage across R_2. Let's assume the normal circuit values are V = 10V, R_1 = 1kΩ and R_2 = 1kΩ.

Figure 3-5. Meter Loading When Making Voltage Measurements

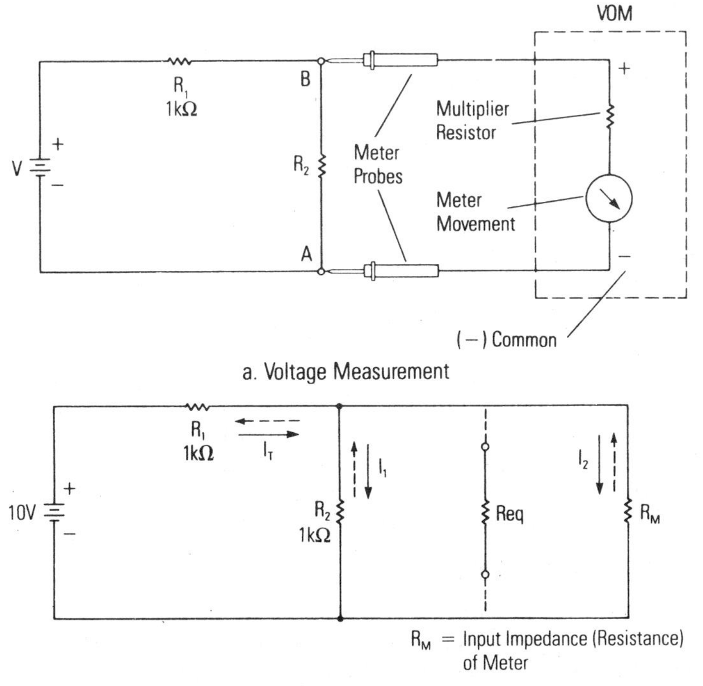

Table 3-1 lists the circuit values and errors when the measurement of V_{R2} is made on the 5-volt scale with VOMs that have input impedances of 200, 1000, 2000, and 20,000 ohms/volt.

Table 3-1. Percent Error Due to R_M

VOM Ohm/Volt	VOM R_M	R_2	Req	R_1	R_T	I_T	V_{R1}	V_{REQ} (V_{R2})	% Error
k	kΩ	kΩ	kΩ	kΩ	kΩ	mA	V	V	%
0.2	1	1	0.5	1	1.500	6.67	6.67	3.33	33.0
1	5	1	0.883	1	1.833	5.45	5.45	4.55	9.0
2	10	1	0.910	1	1.910	5.24	5.24	4.76	4.8
20	100	1	0.990	1	1.990	5.03	5.03	4.97	0.6

When R_M equals 5 kilohms, the voltage measurement is still in error by 9%. Not until a VOM is used that has an internal impedance, R_M, of 10 kilohms (10 times the value of R_2, the resistance across which the voltage is measured) does the error (4.8%) come down into the range of measurement error of the VOM itself (usually 3% to 4%). Therefore, *to prevent meter loading on a voltage measurement, make sure the input impedance (internal resistance in this case) of the voltmeter used for measurement is at least 10 times the impedance (or resistance) across which the measurement is made.*

Accuracy Compromise

The value of R_M, the VOM input impedance, increases as the voltage full-scale range is increased. A 1,000 ohms-per-volt meter has an $R_M = 10{,}000$ ohms on the 10-volt scale and an $R_M = 50{,}000$ ohms on the 50-volt scale. To have R_M large so it does not load the circuit, the full-scale range should be large so the loading error is small. However, to have the greatest reading accuracy, the full-scale range should be as small as possible to make the meter deflection as close to full scale as possible. These are contradictory statements, so the choice of which full-scale range to use is a compromise when the loading of R_M will cause significant error.

Measurement With A DMM

As explained in Chapter 2, electronic circuitry is used in a digital voltmeter. As a result, the input impedance is quite high — usually 1 megohm (one million ohms) or more. Therefore, normal voltage measurements across components that have as high a resistance value as 100,000 ohms will have little or no error. *This is one of the significant advantages of using a DMM — little circuit loading occurs unless resistances approach megohm values.*

BASIC DC MEASUREMENTS

A typical VOM is shown in *Figure 3-6*. It is an instrument that is easy to use and simple to operate; therefore, it is very popular with the handyman or the beginner in electricity and electronics. It functions as a voltmeter, ammeter, or ohmmeter depending on the position of a selector switch. It has 41 ranges. The major ones are: 12 for dc voltage from 0-1,000 volts, 8 for ac voltages from 0-

1,000 volts, 8 for dc current up to a maximum of 500mA and 5 to measure resistance from 0 to 20 megohms (20×10^6 ohms). A special switch doubles the number of ranges on dc voltage and current and ac voltage by cutting the full-scale of each range in half to provide extra resolution and higher accuracy in the reading. The sensitivity of this meter is 50,000 ohms per volt on dc and 10,000 ohms per volt on ac. The sensitivity reduces in half when the scale is cut in half. That means that the meter movement deflects to full scale on dc with just 20 microamperes (0.00002A) of current.

Figure 3-6. VOM

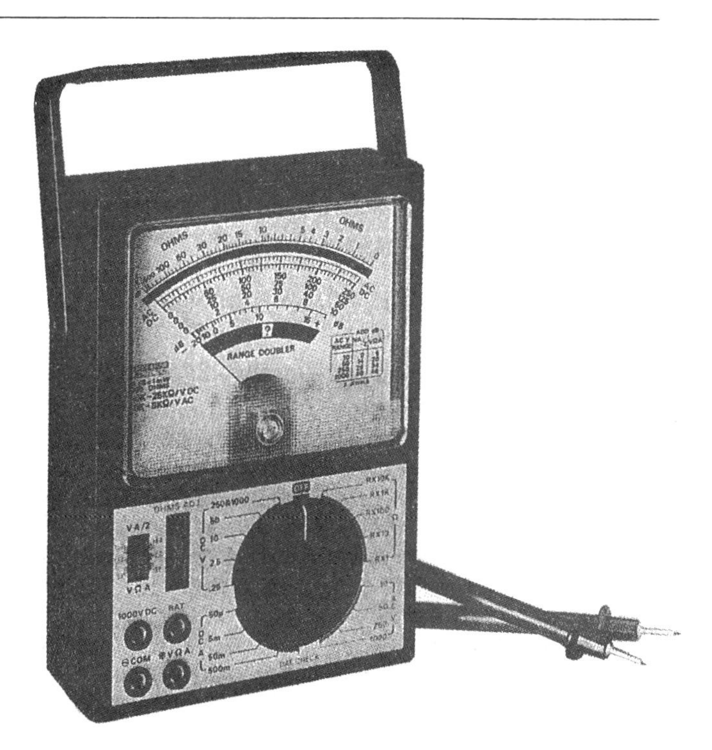

Making a VOM DC Voltage Measurement

To make a voltage measurement, follow the steps listed. The voltage to be measured is the dc voltage across R_3 in *Figure 3-7a.*

1. To prevent an error being introduced, the zero of the meter should be checked before any measurement is made. With a VOM, the instrument should be off and the pointer zero position adjusted mechanically with the zero adjust screw as described in the instrument manual.
2. This is a dc voltage measurement, so set the range selector switch to the dc voltage range desired. Usually, it is best to start with a range greater than the final one. If you do not know the approximate value of the voltage, place the selector switch on the highest range.
3. Place the minus (−) meter probe at point A. Now touch the plus (+) probe momentarily to point B. Watch the meter needle and see if it moves up scale. If no movement is detected, rotate the range switch to the next lowest scale until a deflection is noted.

Figure 3-7. DC Voltage Measurement

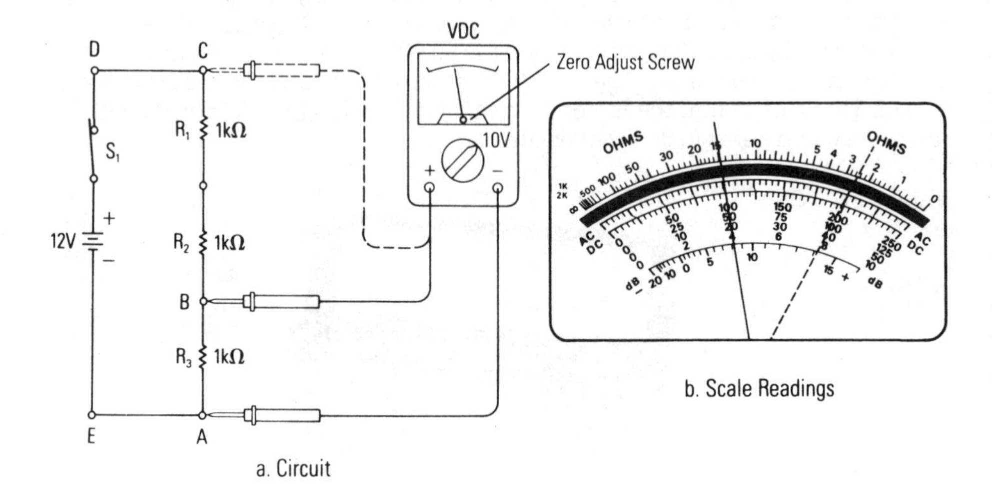

b. Scale Readings

a. Circuit

4. If the deflection is up scale, hold the plus probe to point B permanently. (If your test probes have clips on them, clip the leads in place.)
5. If the deflection is down scale (below zero) reverse the leads or, if the meter has a polarity switch, switch the polarity so that the deflection is up scale.
6. Rotate the selector switch to get the greatest on-scale needle deflection without exceeding the full-scale. In the example of *Figure 3-7*, the voltage is 4 volts; therefore, the selector switch would be set on 10 volts and the meter would read 4 on the 10-volt scale as shown in *Figure 3-7b*. If the 2.5-volt scale were used, the meter needle would "peg" on the right-hand side.
7. The meter of *Figure 3-6* has a switch that reduces the range selected in half. Switching the switch to V/2 reduces the full-scale range to 5 volts and the needle would now read 4 volts as shown in the dotted line position of *Figure 3-7b*. (The actual reading is taken using the 50-volt scale and, because the range selector is on the 5-volt range, every meter reading is divided by 10.) This provides a deflection closer to full scale, and thus, a more accurate reading.
8. Parallax is reduced by aligning the needle with the image in the mirror above the voltage scales.
9. If the battery voltage is to be measured, the plus probe would be momentarily touched to point C of *Figure 3-7a* with the range selector switch on the 50-volt range. Steps 4 through 8 would then be followed to read 12 volts on the 25-volt range.

Making a VOM DC Current Measurement

Current measurements are different. In general, the meter must be inserted into the circuit so that the circuit current is through the ammeter (an exception is the clamp-on ammeter which will be covered in detail later). *Figure 3-8a* shows a VOM used as an ammeter inserted in the series circuit of *Figure 3-7a* to measure the series circuit current. Follow the steps listed for this measurement:

1. With the meter out of the circuit, check the zero position of the meter and mechanically adjust it with the zero adjust screw if required.
2. Turn off the circuit power and disconnect the circuit so that the meter may be inserted. In *Figure 3-8a*, point C is disconnected from point D and the ammeter is inserted. (Remember that the current in a series circuit is the same throughout the circuit, so the ammeter can be inserted at any convenient point.)
3. Set the range selector switch to the highest dc current range. Some VOMs also have a function switch (volts, amperes or ohms). If so, select the dc amperes function.
4. Current will be from the most positive terminal through the meter to a more negative terminal. Therefore, connect the plus probe of the meter to the most positive voltage point (point D).
5. Turn on the circuit power. Touch the minus probe momentarily to point C and observe the meter needle deflection. If the deflection is down scale, reverse the meter leads. Connect both probes securely.

Figure 3-8. DC Current Measurement

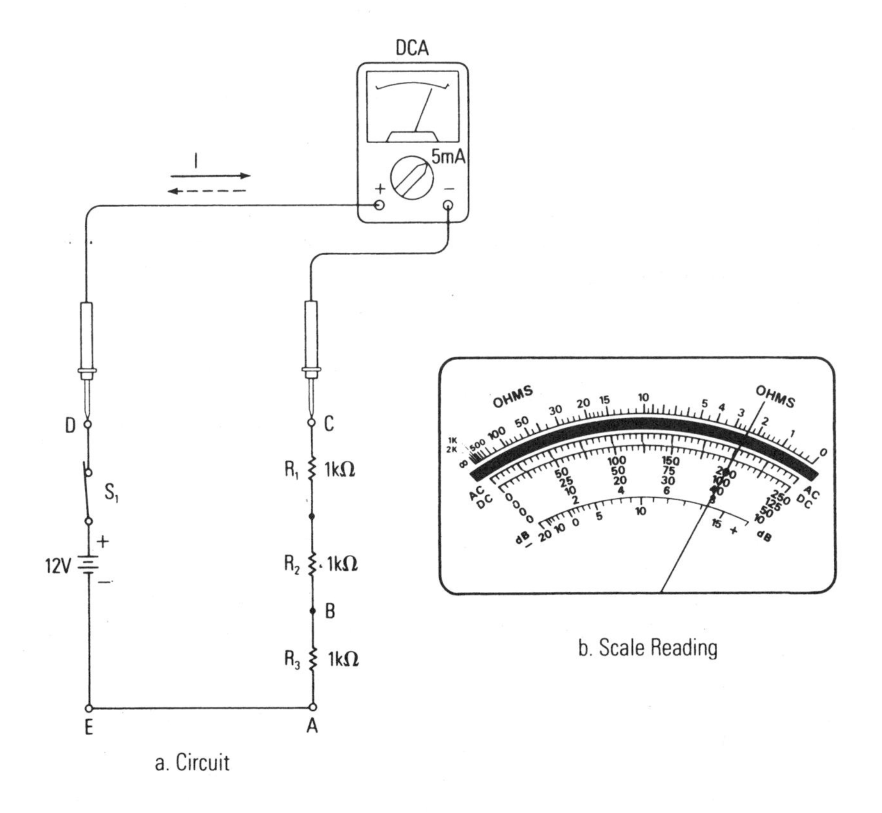

a. Circuit

b. Scale Reading

6. Adjust the range switch to give the greatest on-scale deflection. Since the current in *Figure 3-8a* is 4 milliamperes, the range selector would be set on 5 milliamperes full scale (abbreviated 5mA), and the needle would read 4 milliamperes as shown in *Figure 3-8b.* As for the voltage measurement, the 50 scale is used and all readings divided by 10. There is no need to use the half-scale function switch in this case.

Be aware that the highest current range is 500mA. Some meters have a high range of 10 amperes. If so, the plus (+) lead must be plugged into an auxiliary jack to activate the 10A current range.

Making a VOM Resistance Measurement

Complete information on the condition of electrical circuits and the circuit components cannot be obtained by the use of the voltmeter and the ammeter alone. Using an ohmmeter adds significantly to your information about a circuit. Knowing for sure that a circuit is complete (there is continuity) and being able to measure the resistance of the various circuit components are valuable pieces of circuit information obtained by using an ohmmeter.

Recall that a VOM when used as an ohmmeter supplies its own current from internal batteries to operate the meter movement. *All circuit power must be turned off or removed when making continuity and resistance measurements to prevent damaging or completely burning out the VOM.*

Example Measurement

The meter of *Figure 3-6* uses a series circuit for the ohmmeter. As a result, zero ohms on the resistance scale is full-scale deflection, and open circuit or "infinity" ohms is the idle position of the needle. We will measure the resistance of the circuit of *Figure 3-7a* as shown in *Figure 3-9a.* Follow the steps listed:

1. Make sure power is disconnected from the circuit.
2. Make sure the VOM is set for the ohmmeter (Ω) function. With the ZERO ADJUST screw, mechanically adjust the idle position of the needle so it lines up with the zeros on the ac and dc scales. This is infinity on the OHMS scale.
3. Short the meter leads together and adjust the OHMS ADJUST control to obtain a meter reading of 0 on the OHMS scale. *If this adjustment cannot be made, the batteries in the meter need to be replaced.*
4. Place the meter leads across the resistance to be measured; for example, point A and point C in *Figure 3-9a.*
5. Change the range selector switch to cause the meter reading to be between half-scale and full-scale if possible, or as close to half-scale as possible. If the meter reading is close to the high-resistance end of the scale, select a higher resistance range; if the reading is closer to the low-resistance end, select a lower resistance range. Of course, with high resistances near the maximum that the meter can measure, the meter reading will always be on the high-resistance end of the scale.
6. Since the resistance in this case is 3 kilohms, the range selector would end up on the R × 1K range and the meter would read 3 as shown in *Figure 3-9b.* The range switch selects different values that multiply the meter reading. On the R × 1 scale, the meter reading is taken directly. On the R × 10K, all readings are multiplied by 10,000.

Figure 3-9. Resistance Measurement

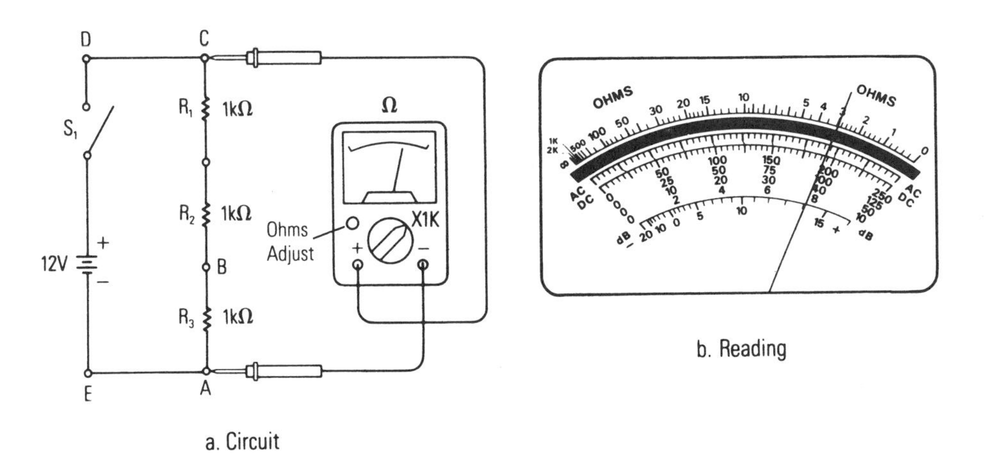

a. Circuit

b. Reading

When making in-circuit measurements, remember that circuit paths in parallel with the component being measured may cause reading errors. Check the circuit diagram for the presence of such parallel paths before assuming that the reading obtained is correct.

DMM DC Voltage Measurements

A digital multitester like the one shown in *Figure 3-10* is chosen for our example measurements. Two significant differences are apparent between this DMM and a VOM. First, the output is a digital display where the meter reading is a particular number. Second, since electronic circuits are required to produce the conversion required, power must be supplied to the electronic circuits; therefore, the DMM has a power switch which must be ON for all measurements.

As shown in *Figure 3-10,* beyond these differences, the DMM and VOM are functionally very similar. A range selector switch selects the function and the range, the meter leads are plugged into a common (−) jack and a plus (+) jack (V-Ω-mA) to make measurements, or the plus lead is plugged into special auxiliary jacks for special current ranges (10A in this case).

The steps for making a voltage measurement with a DMM in the circuit of *Figure 3-7* are the same as they are when a VOM is used. The exceptions are that no zero adjustments or polarity precautions are necessary because the meter takes care of these automatically. The actual reading will be a number on the digital display. For example, measuring the voltage drop from point A to point B in *Figure 3-7a* would produce a display of 4.00 on the 20-volt range.

Polarity

Test leads need not be switched when measuring with a DMM. The display's polarity indicator displays whether the voltage is positive or negative with respect to the common lead. No polarity indication means the polarity is positive.

Figure 3-10. DMM Used for Example Measurements

Overrange and Fused Protection

The three lower-order (least significant) digits of the display are blanked, or the most significant digit may flash, if the quantity measured is overrange. On other DMMs, there may be a special overrange indicator that appears on the display to indicate that the reading is greater than the selected full-scale range. Change the range selector to the next highest range until a digital reading is obtained or the highest range is reached. The voltage being measured exceeds the highest full-scale reading if an overrange indication is obtained on the highest range. Many instruments also are fused to protect against excessive current on any range.

Low Battery Voltage

A special indicator (LO BAT for meter in *Figure 3-10*) usually is displayed to indicate when the internal batteries need to be replaced.

DMM DC Current Measurements

The current measurement functionally is the same as the steps outlined for the VOM current measurement, except, as for the voltage measurements, the leads need not be reversed for polarity, and there is no zero adjustment. A measured value of 4 milliamperes would be indicated as 4.0 on the display if the 200 milliamperes range is selected. Recall that the actual method of measuring the current is explained in Chapter 2.

DMM Resistance Measurements

The same precautions must be followed when measuring resistance with a DMM as with a VOM. Remove power from the circuit before making resistance measurement. The measurement is made with the test leads across the component to be measured. Current from the DMM causes a voltage drop across the component resistance and the voltage is read by the DMM and converted to a direct resistance measurement.

When measuring the resistance of *Figure 3-9,* the DMM would display 3.00 on the 20-kilohm range if the resistance is exactly 3000 ohms.

BASIC AC MEASUREMENTS

Here are some factors that are important when making ac measurements.

1. AC voltage and current measurements depend on frequency. Consult the meter manual to determine the limits of the meter's frequency response.
2. Most VOMs read in RMS values, but ac meters can either read average, RMS, or true RMS values. *Only true RMS reading meters do not require the ac signal to be a sine wave without producing error.*
3. AC current measurements are accomplished by rectifying a voltage developed across a resistor in the circuit.
4. AC VOM readings are accurate to a percent of full-scale, but the accuracy and sensitivity are usually less than for dc measurements.
5. Non-linear rectifier characteristics make small ac voltage measurements less accurate.
6. Making an ac measurement with a VOM usually has more loading effect on the circuit than when making a dc measurement.
7. A DMM usually has the same accuracy, sensitivity and circuit loading for ac as it does for dc.
8. When an ac voltage is applied to circuits that have only resistance (such as *Figures 3-7* and *3-8*), the measurement steps are the same as outlined for dc measurements. Of course, there is no polarity consideration for ac measurements.

AC Impedance and Reactance

The opposition to current in a dc circuit is termed resistance. If the opposition to current in an ac circuit is due to resistance, the effect is the same; however, if the circuit contains inductors or capacitors, the opposition to ac current is more complex. Inductors are coils of wire wound around a core, usually made of iron. They are contained in motors, generators, solenoids, relays, chokes, and transformers. Capacitors store charge on parallel conductive plates which are separated by a dielectric (insulator). Direct current cannot pass through a capacitor. Inductors store energy in magnetic fields; capacitors store energy in electrostatic fields.

Reactance

The opposition offered by an inductor or capacitor is called reactance. The symbol for reactance is X and the unit is ohms. The reactance of each of the components is frequency sensitive. As explained in the Appendix, inductive reactance is expressed as:

$$X_L = 2\pi fL$$

and capacitive reactance is expressed as

$$X_C = \frac{1}{2\pi fC}$$

where f is frequency in hertz (cycles per second), L is inductance in henries, C is capacitance in farads, and π is the constant 3.1416.

If f = 0 (which is dc), X_L is zero and X_C is infinity.
If f = infinity, X_L is infinity and X_C is zero.

Impedance

Impedance is the term given to the total opposition to ac current when both resistance and reactance are present to impede the current. The reactance can be inductive, capacitive or a combination of both. What makes impedance and ac voltage and current complex is that measured quantities do not add directly. They must be added vectorially.

Vector Addition

How voltage, current and impedance add vectorially is best demonstrated by an example. Look at *Figure 3-11.* A 60Hz, 50VAC power source is connected to a resistor and solenoid in series. The resistor has 30 ohms resistance and the solenoid has 40 ohms inductive reactance. To find the total impedance of the circuit, the resistance and reactance must be combined at right angles. As shown in *Figure 3-11,* the resistance is plotted horizontally and the reactance is plotted 90° upright from it, joined tip-to-tail and represented by arrows, called vectors, with appropriate length and to the same scale. The total impedance, Z, is another vector from the start (tail) of the resistance to the tip of the reactance. This forms the hypotenuse (the long side) of a right triangle.

A mathematical theorem (Pythagorean's Theorem) states that the hypotenuse of a right triangle is equal to the square root of the sum of the squares of the sides. Therefore, the total impedance of the series circuit and the total current can be calculated as

$$Z = \sqrt{(30^2 + 40^2)}$$
$$Z = \sqrt{(900 + 1600)}$$
$$Z = \sqrt{2500}$$
$$Z = 50\Omega$$

$$I = \frac{V}{Z}$$
$$I = \frac{50V}{50\Omega}$$
$$I = 1A$$

The total current is one ampere.

Ohm's law for ac circuits can now be used to find the voltage across the respective components. The voltages across the resistance and inductance are:

$$V_R = I \times R$$
$$V_R = 1A \times 30\Omega$$
$$V_R = 30V$$

$$V_L = I \times X_L$$
$$V_L = 1A \times 40\Omega$$
$$V_L = 40V$$

Voltage Measurements

If a VOM on an ac voltage range of 50 volts is used to measure the voltage across the resistance and inductance, the voltages measured would be 30 volts across the resistance and 40 volts across the solenoid. If now the source voltage is measured, it would measure 50 volts. Each of these measurements is shown in *Figure 3-11.* Note that the voltages across each of the components add directly to 70 volts, but the source voltage is only 50 volts. In circuits where the total impedance is made up of resistance and reactance, the measured voltages must be added vectorially. *Figure 3-11* shows how the total voltage of 50 volts is the vectorial sum of the 30 volts across the resistance and 40 volts across the solenoid. The steps for making the voltage measurements are basically the same as for dc.

Current Measurements

Current measurements are made in ac circuits by using the current function on the selector switch and inserting the meter leads into the circuit just as with dc current measurements. If your meter does not have an ac current function, current measurements can be made easily by inserting a small known value of resistance into the circuit and measuring the ac voltage across it. An example is shown in the dotted insert of *Figure 3-11*. The voltage drop across the one ohm resistor used for the current measurements may be disregarded when the voltage measurements are made.

CAUTION:

When making current measurements in ac circuits that contain parallel branches with inductors and capacitors in parallel, very high circulating currents can be encountered in the parallel branches at particular frequencies of the applied voltage.

Figure 3-11. Voltage Measurements of Circuit with Reactance

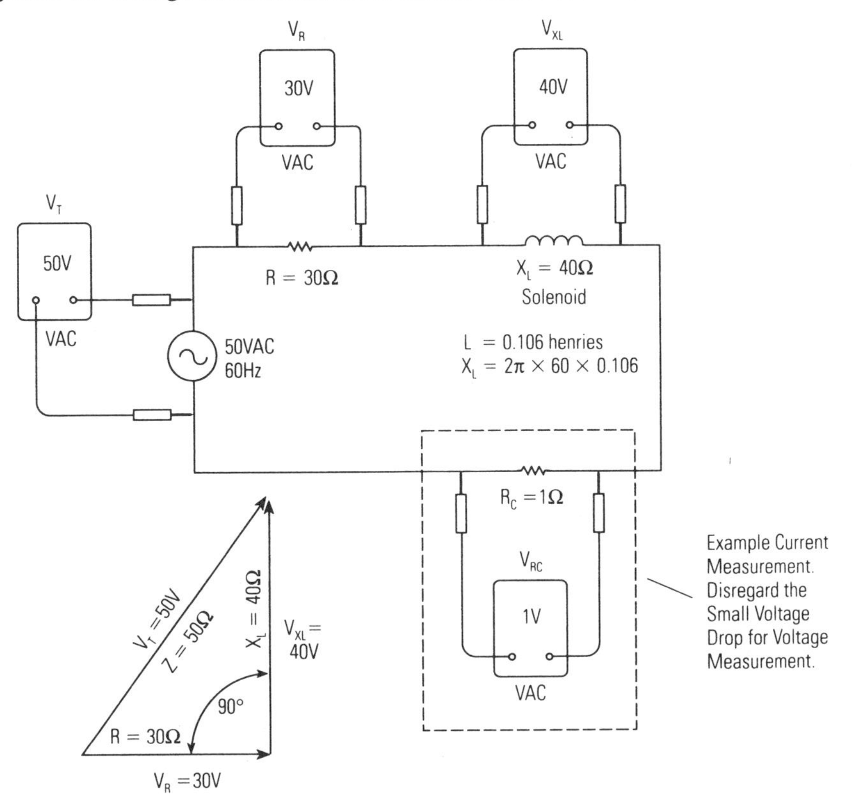

Impedance Measurements

Even though there are special impedance meters where meter scales are calibrated in ohms impedance, VOMs and DMMs normally do not have impedance scales as they have resistance scales. Impedances are obtained by measuring voltage and current and calculating impedance.

CLAMP-ON CURRENT METERS

Current measurements may be made more easily and without contacting the circuit by using a clamp-on current meter. These instruments also help protect the user from potentially dangerous voltages in high-power circuits. *Figure 3-12* shows examples of clamp-on meters. One is a DMM and the other is a VOM.

Figure 3-12. Clamp-On Ammeters

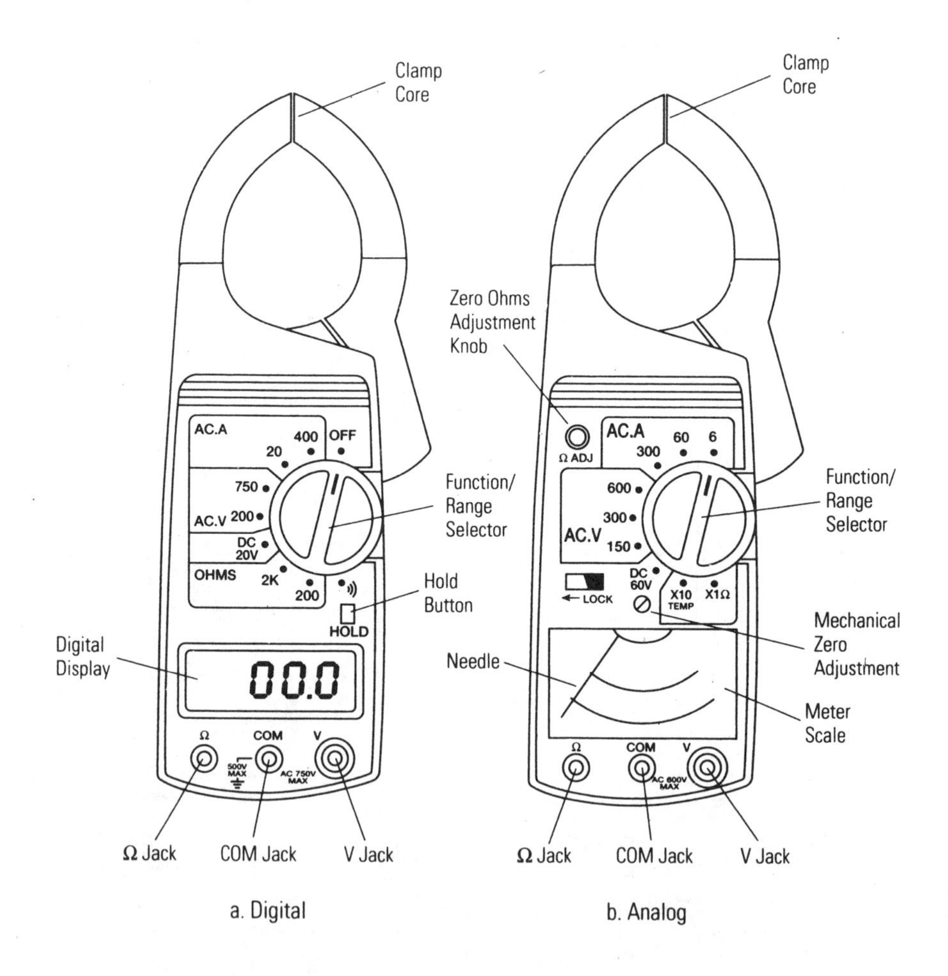

Most VOMs and DMMs can measure current up to only 10 amperes. Larger currents can be measured more safely and conveniently by using a clamp-on type. There are two types of pickups that may be used on clamp-on meters: inductive and Hall-effect. The inductive type is used on meters that measures ac current only. Hall-effect type pickups can measure both ac or dc current. The clamp-on technique can be used with either an analog or digital meter. As an example, the digital clamp-on meter shown in *Figure 3-12a* measures ac current up to 400A, ac voltage to 750V, dc voltage to 20V, resistance to 2kΩ, has a continuity buzzer, and has a data hold button. *Figure 3-13* shows a clamp-on meter measuring the current to a load.

The Hall-Effect

A Hall-effect device is a semiconductor device that produces an output voltage that is proportional to the magnetic field (changing or steady) in which it is placed. The important difference between the inductive type pickup and the Hall-effect one is that the magnetic field must be changing for the inductive type to produce an output. The Hall-effect pickup produces an output even with a steady magnetic field. The principle of the Hall-effect is shown in *Figure 3-14*. A small current must be maintained through the semiconductor material (either "P" or "N" type may be used). The magnetic field deflects the movement of charge through the crystal and causes a difference of potential across its sides.

Some VOMs today have a clamp-on accessory that may be used to provide a clamp-on ammeter for measuring ac currents. The clamp-on adapter is clamped around a wire carrying a large ac current and the current is then transformed to an ac voltage. The VOM, with special current scales, is used as an ac voltmeter to measure the ac current in the conductor.

Figure 3-13. Clamp-On Ammeter Measuring Current to a Load

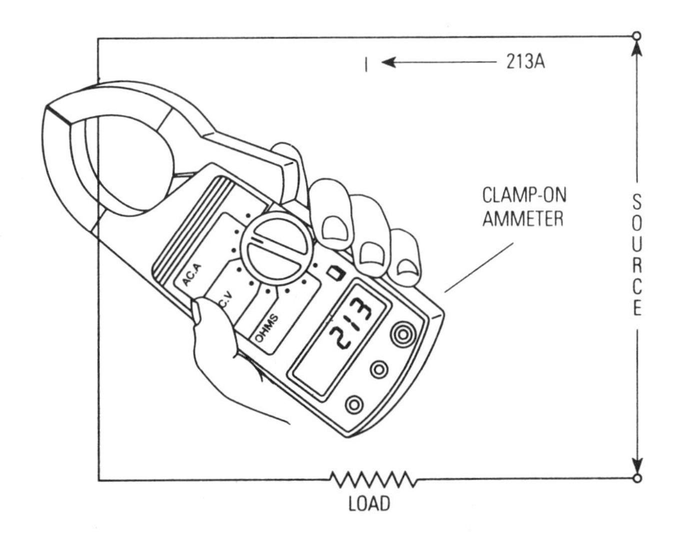

Figure 3-14. Principle of the Hall-Effect

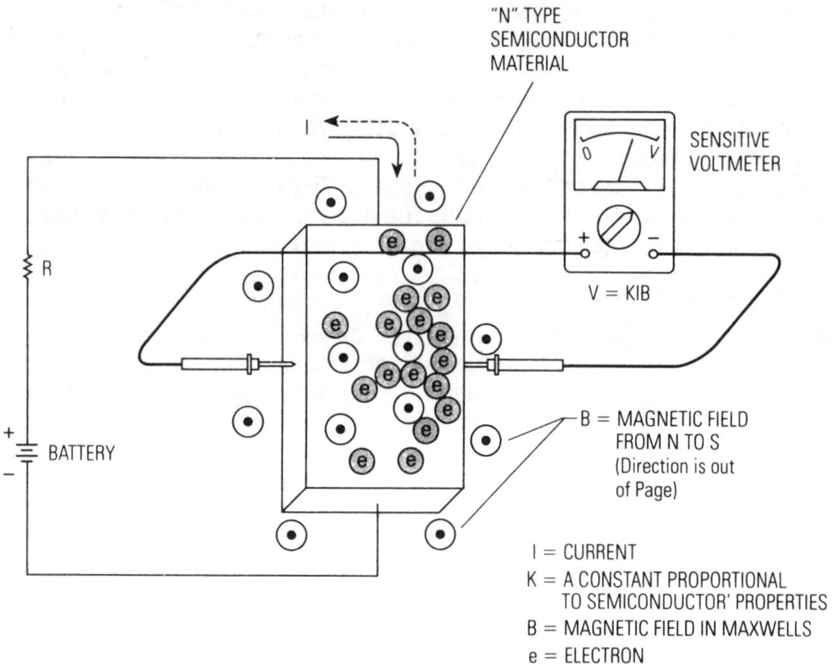

"N" TYPE SEMICONDUCTOR MATERIAL

SENSITIVE VOLTMETER

V = KIB

R

BATTERY

B = MAGNETIC FIELD FROM N TO S (Direction is out of Page)

I = CURRENT
K = A CONSTANT PROPORTIONAL TO SEMICONDUCTOR' PROPERTIES
B = MAGNETIC FIELD IN MAXWELLS
e = ELECTRON

POWER MEASUREMENTS

Indirect measurement of voltage and current will provide the power being delivered or dissipated in a circuit. The power P, using a measured voltage V, and current I is calculated by:

$$P = I \times V$$

The clamp-on meter described above is an excellent way to obtain the current measurement for the power calculation.

DC Circuits

In dc circuits the calculation is a direct one. Multiply the voltage measured by the current measured. The power is dissipated as heat.

AC Circuits

In ac circuits, reactance again complicates the power calculation. The multiplication of voltage times current may not result in true dissipated power, but may result in an "apparent" power because of the reactance present. Consult a more detailed text on ac circuits for a more complete understanding of the subject.

SUMMARY

With the description of the meters and how they operate, and a basic understanding of measurements, we move on in the next chapter to some actual practical examples of meter measurements.

CHAPTER 4

Measuring Individual Components

MEASURING INDIVIDUAL COMPONENTS

An important part of the field of electronics is understanding the various components and electron devices that make up complex electronic circuits. Every electrical circuit has three inherent properties: resistance, capacitance and inductance. Using a multitester to measure these properties of common components is the subject of this chapter. Details expand on the initial measurements made in Chapters 1 and 2.

HOW TO MEASURE A RESISTOR

Resistor Basics – Fixed and Variable

Resistance in a circuit is the opposition to current. All materials used in an electric circuit will offer some resistance. Those offering low resistance are called conductors and are used as paths for current. Those offering extremely high resistance offer no path for current and are called insulators. Components manufactured specifically to be placed in a circuit to provide resistance are called resistors and are used most frequently in electrical circuits to limit current or to create voltage drops. Insulators are used to resist or prevent current in certain paths.

Resistors come in different values, shapes and sizes. Resistors are classified two ways — by resistance measured in ohms and by power rating measured in watts. The power rating of resistors for electronic circuits range from 1/8 watt to hundreds of watts. The ohmic values range from hundredths of ohms (0.01) to hundreds of megohms (100×10^6). Manufacturers have adopted a standard color code system for indicating the resistance or ohmic value of low power resistors (normally below 2 watts). Higher power resistors usually have the resistance value imprinted on their bodies. The color code system is in the Appendix.

Measuring a Fixed Resistor

Out of Circuit

One of the easiest measurements to make with a VOM is to measure the value of a fixed resistor out of a circuit. Use the VOM as an ohmmeter and connect the test leads of the ohmmeter across the resistor as shown in *Figure 4-1.*

Figure 4-1. Measuring a Resistor Out of a Circuit

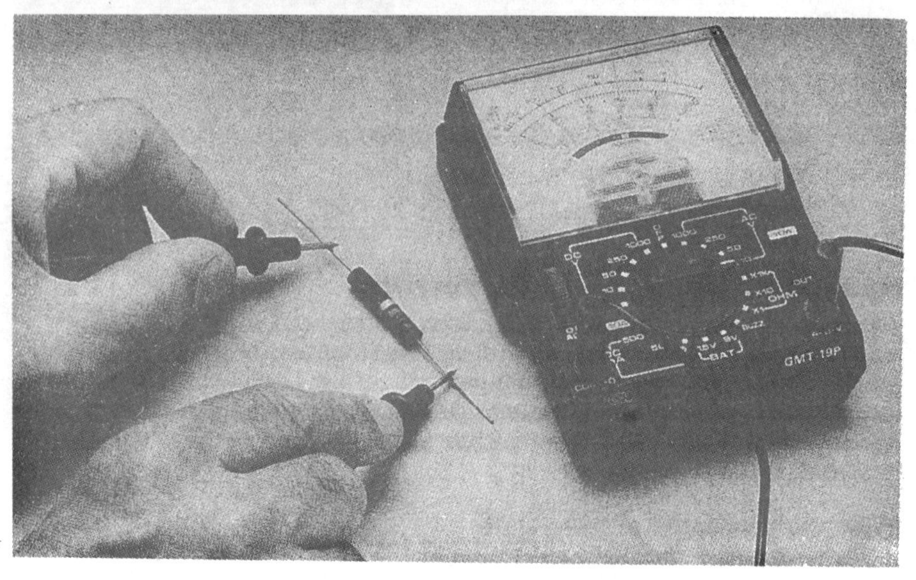

In Circuit

Recall from Chapter 1 the two precautions that must be observed when using an ohmmeter to measure a resistance in a circuit:

1. The power source must be turned off. Disconnect the equipment from the power source, if possible.
2. When the resistor is in a circuit, if possible, disconnect one end of the resistor from any circuit or additional component so only the resistance of the single resistor is measured.

Figure 4-2 shows the correct way to measure the resistance of R_3. Notice the supply V_S has been disconnected. Also, note that one end of R_3 is disconnected from the circuit so the ohmmeter measures the resistance of R_3 alone. If R_3 is not disconnected, then the value of resistance measured would be R_3 in parallel with R_2.

Measuring a Variable Resistor

When a potentiometer is checked with an ohmmeter, the total resistance is first measured from end-to-end as shown for point A in *Figure 4-3*. These components typically have a $\pm 20\%$ tolerance. Next, the resistance should be tested from the wiper arm to one end as the potentiometer is rotated through its full range as shown for Point B in *Figure 4-3*.

The resistance should vary smoothly from near zero to the full value of the resistance. Any sudden jump to either a higher or lower value, or any erratic reading indicates a defective spot on the resistance element. (This is one measurement where a VOM has the edge on the DVM because erratic movement of the meter needle is easily seen.) If the wiper arm is not making firm contact with the resistance element, simply tapping the case of the

Figure 4-2 Measuring a Resistor in a Circuit

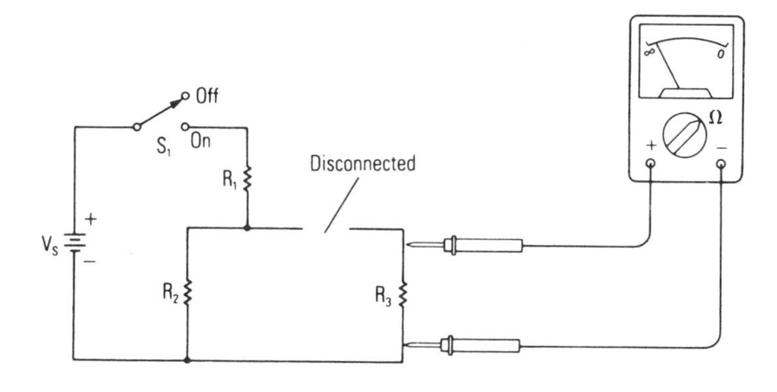

Figure 4-3. Measuring a Variable Resistor

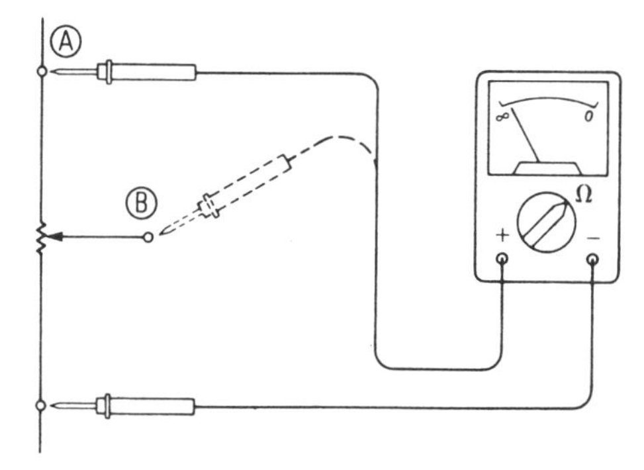

potentiometer may produce erratic resistance readings. Any of these erratic indications mean a defective or dirty potentiometer. A spray cleaner may salvage the unit, however, if, after cleaning, it does not produce a smooth resistance change over its entire range, it should be discarded.

Rheostats may be tested by a similar process with the exception that the only measurement is from the wiper to one end since the rheostat has only two terminals. In general, rheostats will be open if they are defective because they usually are high wattage units and must handle high current.

Thermally Intermittent Resistors

All electronic components, including fixed and variable resistors, can be thermally intermittent. Measure the resistance while subjecting the suspected component to extreme temperature change to detect this type of defect. Electronics parts stores stock an aerosol spray component cooler to spray on a resistor to cool it. The tip of a soldering iron can be used to heat it. Be careful not to apply excessive heat which can damage the resistor. A smoothly varying resistance with temperature change is normal, but a sudden or erratic change in resistance as temperature is changed indicates the resistor is thermally intermittent and defective.

"Shifty" Resistors

There is a class of resistors whose resistances change as the resistors' operating conditions change. Common ones are thermistors, varistors, and photoconductors. Thermistors and photoconductors can be measured with an ohmmeter. The resistance of a varistor is calculated from voltage and current measurements.

Thermistors

A thermistor is a resistor that is specifically designed so its resistance varies with temperature in a certain way. It exhibits large negative temperature characteristics; that is, the resistance decreases as the temperature rises and increases as the temperature falls.

Varistors

A varistor is a resistor whose resistance is voltage dependent. Its resistance decreases as voltage across it is increased.

Photoconductors

A photocell's resistance (photoconductor) varies when light shines on it. When the cell is not illuminated, its "dark" resistance may be greater than 100K ohms. When illuminated, the cell resistance may fall to a few hundred ohms. These values can be measured with an ohmmeter.

HOW TO MEASURE A CAPACITOR

Capacitor Basics

Capacitance is the property whereby two conductors separated by a non-conductor (dielectric material) have the ability to store energy in the form of an electric charge and oppose any change in that charge. The operation of a capacitor depends on the electrostatic field set up between the two oppositely charged parallel plates.

The unit of capacity is the farad, named in honor of Michael Faraday. It is the amount of capacitance which will cause a capacitor to attain a charge of one coulomb when one volt is applied. Expressed as a mathematical equation:

$$C = \frac{Q}{V}$$

where C is one farad when Q is one coulomb and V is one volt.

One farad is too large for practical applications; therefore, smaller values are used. A microfarad (μF) is 10^{-6} farads, a nanofarad (nF) is 10^{-9} farads, and a picofarad (pF) is 10^{-12} farads. Capacitors with values in microfarads and picofarads are very common in electronic circuits.

The physical factors which determine the amount of capacitance a capacitor offers to a circuit are:

a. the type of dielectric material, (K);
b. the area in square meters of the plates, (A);
c. the number of plates, (n); and
d. the spacing of the plates in meters, which also is the thickness of the dielectric, (t).

The basic capacitor and its symbol are shown in *Figure 4-4* with the relationship between the physical factors indicated so the amount of capacitance can be calculated. In *Figure 4-4b,* n = 6, but the most common capacitors have only two parallel plates. A common example is a coupling capacitor used in amplifier circuits. It is made up of two long strips of metal foil separated by a dielectric, rolled up into a small compact size and encapsulated in plastic. Pigtail leads extend out from the plastic body to provide the electrical connection into electronic circuits.

To change a capacitor's charge, and thus its voltage, there must be current in the circuit containing the capacitor. The time it takes to change the charge depends on the capacitance and the impedance of the circuit.

Figure 4-4. Basic Parallel Plate Capacitor

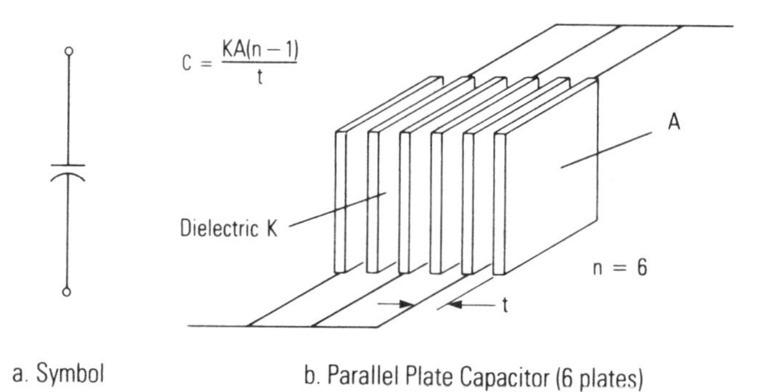

Measuring Capacitors and Semiconductors

When measuring capacitors and semiconductors with an ohmmeter, you need to know the polarity of the voltage supplied through the probes. In most cases the red lead is positive, but sometimes the internal battery negative is connected to the positive (V-Ω-A) lead.

CAUTION:

Before a capacitor is measured with an ohmmeter or DMM capacitance meter, remove it from the circuit and short across its leads or plates to make sure it has no residual charge. Such residual charge could damage an ohmmeter or DMM.

Relative Amount of Capacitance

Two capacitors can be compared as to their relative capacitance by using a VOM. The amount of needle deflection of an ohmmeter can be used to indicate a relative amount of capacitance. By connecting the ohmmeter to the capacitor as shown in *Figure 4-5a,* the ohmmeter battery charges the capacitor to its voltage. The meter will deflect initially and then fall back to infinity as the capacitor charges. In *Figure 4-5b,* after the initial charge of the capacitor, the ohmmeter leads are reversed and the capacitor voltage is now in series with the voltage inside the ohmmeter. The charge on the capacitor is aiding the ohmmeter battery. The needle now deflects a larger deflection proportionally to the amount of capacitance, and then decays as the charge is redistributed on the capacitor.

Figure 4-5. Measuring Relative Amount of Capacitance

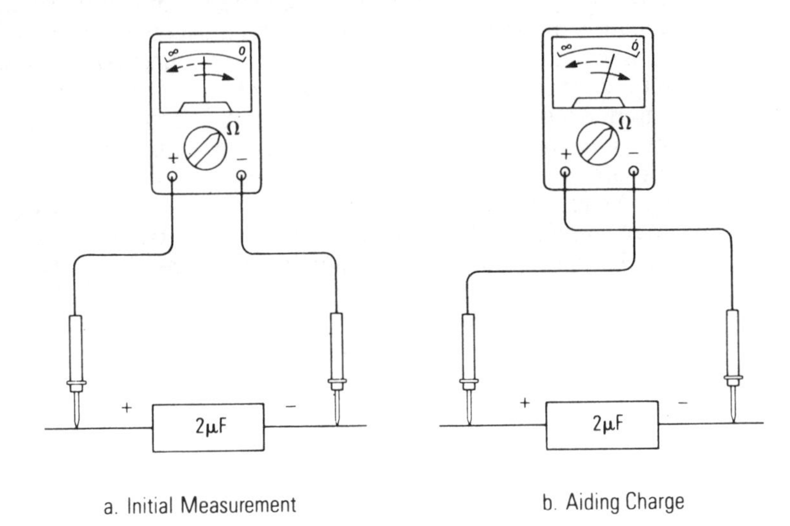

Leaky and Shorted Capacitors

Paper, mica and ceramic capacitors fail in two ways. The dielectric breaks down and the capacitor plates short together, or the capacitor becomes "leaky." When "leaky," the dielectric still supports a voltage but the dielectric resistance becomes much lower than normal. Both of these conditions can be detected with a VOM or DMM. There are two checks that can be made. The first is simply a resistance measurement using the VOM or DMM as an ohmmeter across the terminals of the capacitor. If the capacitor is shorted, the ohmmeter will read zero or a very low value of resistance. If the capacitor has become "leaky," then the resistance measurement will be much less than the normal nearly infinite reading for a good capacitor. Leaky capacitors need to be replaced before they turn into shorted capacitors.

In some capacitors, the dielectric does not become "leaky" until a voltage is applied. That is, it breaks down under load. This defect cannot be detected with an ohmmeter but can be found by using the VOM or DMM as a voltmeter as shown in *Figure 4-6.* A dc voltage is placed across the series combination of the voltmeter and the paper, mica or ceramic capacitor (not for electrolytic capacitors unless proper polarity is maintained). A good capacitor will show only a momentary deflection on the voltmeter, then the reading will decay to zero volts as the capacitor charges to the supply voltage. A defective capacitor will have a low insulation resistance, R_{INS}, (it may be at a particular voltage), and will maintain a voltage reading on the meter. The lower the insulation resistance of the capacitor, the greater value the voltmeter will read. When insulation resistance is checked by this method, it is in series with the meter. Because R_{INS} is normally high, it limits the current; therefore, a change in the VOM voltmeter range does not significantly affect the total resistance of the circuit. The percentage of meter-scale deflection remains fairly constant with different voltmeter ranges. The power supply voltage V_S should be set for the rated working voltage of the capacitor for this test.

Figure 4-6. Measuring R_{INS} Using a Voltmeter

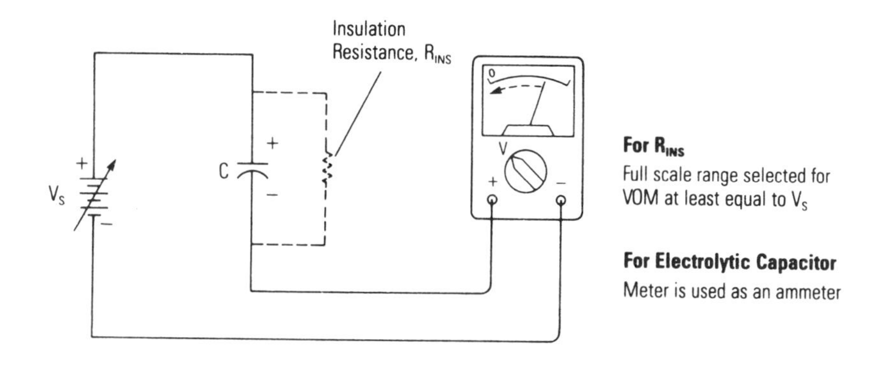

If the insulation resistance is such that it produces a scale reading on a VOM or DMM, the R_{INS} may be calculated by using the following equation:

$$R_{INS} = R_{INPUT} \frac{V_S - V_M}{V_M}$$

where V_S = supply voltage
V_M = VOM or DMM measured voltage
R_{INPUT} = VOM or DMM input resistance
R_{INS} = capacitor insulation resistance

Measuring an Electrolytic Capacitor

Special care must be taken when measuring electrolytic capacitors because they are polarized. As a result, when using a VOM as an ohmmeter to test an electrolytic capacitor, the ohmmeter test lead polarity must be correct to give the proper indication. The test lead arrangement that gives the highest resistance reading for a capacitor is the one to use.

Leakage Current of Electrolytic Capacitors

Measuring the leakage current of an electrolytic capacitor is the best way to judge whether the capacitor is still useful. The circuit of *Figure 4-6* is used for measuring leakage current of electrolytic capacitors except now the meter is an ammeter. V_S, the capacitor's rated voltage, is applied across the capacitor, and the leakage current is indicated by the series ammeter. The maximum permissible leakage current of a new electrolytic capacitor is related to the voltage rating and capacitance of the capacitor according to the following equation:

$$I = kC + 0.30$$

where I = leakage current in mA
C = rated value of capacitor in microfarads
k = a constant as given in *Table 4-1*

Table 4-1. Electrolytic Capacitor Constant K

k	WVDC, volts
0.010	3 – 100
0.020	101 – 250
0.025	251 – 350
0.040	351 – 500

Many factors affect the amount of leakage current. Examples are: the age of the capacitor, how near its rated voltage it has been working, how long it has been uncharged in a circuit, or how long a new one has been on the shelf. If the capacitor exceeds the permissible leakage values, it should be discarded. Experience will help make this test more conclusive.

HOW TO MEASURE INDUCTORS

The property of an electric circuit or component that opposes changes in circuit current is called inductance. The ability of the circuit or component to oppose changes in current is due to its ability to store and release energy that it has stored in a magnetic field. Every circuit has some inherent inductance, but devices which purposely introduce inductance to a circuit are called inductors. Some basics of inductance are given in the Appendix. Tests on inductors with a VOM or DMM are somewhat limited to continuity and shorts.

Continuity and Shorts

Inductors become defective because insulation breaks down and turns short together or the coil shorts to the core. Simple continuity checks with an ohmmeter between one end of the coil and the other, and one end of the coil and the core detect continuity of the coil and if the coil is shorted to the core. A few shorted turns on an inductor are very difficult to detect. If one-half the coil shorts out, a resistance check should detect it, but very accurate measurements must be made to detect a few shorted turns.

TRANSFORMER BASICS

As shown in *Figure 4-7*, a transformer is a device for coupling ac power from a source to a load. A conventional transformer consists of two or more windings on a core that are insulated from each other. Energy is coupled from one winding to another by a changing magnetic field. An ac voltage applied across the primary causes primary current. The changing current sets up an expanding and collapsing magnetic field which cuts the turns of the secondary winding. This changing magnetic field induces an ac voltage in the secondary which produces a current in any load connected across it.

The primary and secondary are wound on an iron core for low frequencies, as in the case of power and audio transformers. In higher frequency circuits, the primary and secondary windings often are wound on an air core.

Ideal Transformer

If the transformer were ideal, there would be no power loss from primary to secondary and 100% of the source power would be delivered to the load. Since voltage multiplied by current equals power, the power relationship is given by:

$$V_P \times I_P = V_S \times I_S$$

where
V_P = primary voltage in volts
I_P = primary current in amperes
V_S = secondary voltage in volts
I_S = secondary current in amperes

In an ideal transformer, the ratio of the voltage induced in the secondary, V_S, to the primary voltage, V_P, is the same as the ratio of the number of turns in the secondary, N_S, to the number of turns in the primary, N_P. The following equation expresses the relationship:

$$\frac{V_S}{V_P} = \frac{N_S}{N_P}$$

The turns ratio of a transformer is the ratio of the secondary turns to the primary turns:

$$\frac{N_S}{N_P}$$

and the secondary voltage is the turns ratio times the primary voltage:

$$V_S = \frac{N_S}{N_P} V_P$$

Figure 4-7 Simple Transformer

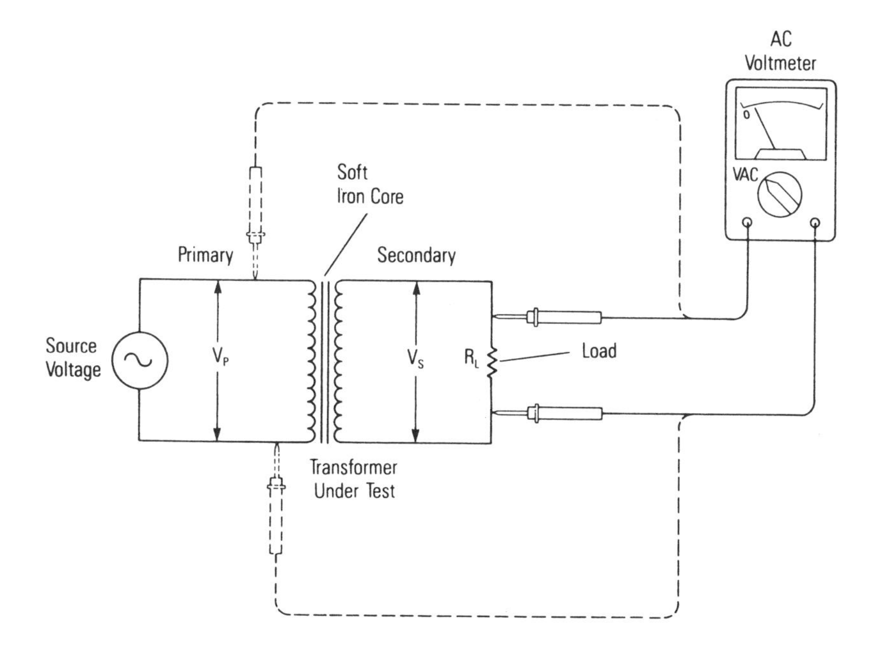

Measuring Turns Ratio

The turns ratio of a transformer can be measured using the circuit of *Figure 4-7* except that the load should be removed from the secondary. Apply a small ac voltage to the primary (V_P in *Figure 4-7*). A good source for this is a doorbell transformer. It supplies from 12 to 18 volts ac. Measure V_P with an ac voltmeter as shown in *Figure 4-7*. Now measure the secondary voltage V_S with the same voltmeter. The turns ratio:

$$\frac{N_S}{N_P}$$

is equal to the secondary voltage divided by the primary voltage:

$$\frac{V_S}{V_P}$$

Step-Up and Step-Down Transformers

If the number of turns in the primary and the secondary are equal, then the voltages appearing across the primary and secondary are equal. This type of transformer, with a one-to-one turns ratio, is called an isolation transformer. If a lower voltage appears across the secondary than across the primary, it is called a step-down transformer. The turns ratio would be less than 1. However, if a higher voltage appears across the secondary than across the primary, it is called a step-up transformer. The turns ratio would be greater than 1. According to the primary and secondary power relationship equation given previously; if the primary voltage and current are held constant, the secondary current will decrease if the secondary voltage is stepped up, and the secondary current will increase if the secondary voltage is stepped down.

Step-Up Transformer Measurement

For example, if the applied V_P is 12 volts and V_S is measured as 60 volts, then the turns ratio is calculated as follows:

$$\frac{N_S}{N_P} = \frac{V_S}{V_P} = \frac{60V}{12V} = \frac{5}{1}$$

The turns ratio is 5:1.

Step-Down Transformer Measurement

If the 110 VAC line voltage is within the rating of the transformer's primary and it is known that the transformer has a step-down secondary, then the 110 volts can be applied to the primary and the stepped-down voltage read with the voltmeter. If V_S is 10 volts, then the turns ratio is calculated as follows:

$$\frac{N_S}{N_P} = \frac{V_S}{V_P} = \frac{10V}{110V} = \frac{1}{11}$$

The turns ratio is 1:11.

If the transformer's primary rating is unknown or if it is unknown whether the transformer is step-up or step-down, it may be dangerous to apply 115 volts to the primary winding. The transformer could be damaged or the secondary voltage could be dangerously high if it has a step-up turns ratio. Thus, applying a lower voltage to the primary for the turns ratio measurement is much safer.

Resistance Testing of Transformer Windings

Continuity

Resistance test with an ohmmeter may be made on most small transformers that are used in electronics to determine the continuity of each winding. Comparison of the measured resistance with the published data from the manufacturer should determine if a suspected transformer is defective. Power transformers and audio output transformers usually have their windings color coded so that the respective winding can be measured with an ohmmeter to determine if there is continuity, and to measure the winding resistance. If a winding measures infinite resistance, the winding is "open." The break may occur at the beginning or end of the winding where the connections are made to the terminal leads. This type of break may be repaired by resoldering the leads to the winding. If the discontinuity is deeper in the transformer, the transformer will have to be replaced.

If the winding resistance is "very" high compared to its rated value, there may be a cold solder joint at the terminal connections. If the condition cannot be corrected, the transformer will have to be replaced.

Shorts – Primary and Secondary

A short from a winding to the core or to another winding may be found by measuring the resistance on a high ohms scale from core to the winding or from winding to winding. Place the ohmmeter leads on the winding lead and on the core, or across from one winding lead to another winding lead. Any continuity reading at all indicates leakage to the winding from the core or between windings; in either case, the transformer is defective. A few shorted turns are difficult to detect, but if a large percentage of the transformer is shorted out, resistance measurements will detect it. The winding-to-core or winding-to-winding resistance of a transformer or an inductor can be tested with a voltmeter and a dc power supply as shown in *Figure 4-8*. The voltmeter will read zero if there is no breakdown between windings and core. If there is significant voltage read on the voltmeter, then there is a significant reduction in the interwinding resistance and the transformer is going bad.

MEASURING SEMICONDUCTOR DEVICES

The only definite test for a transistor is operating in the circuit for which it was designed. However, there are several tests that can indicate the condition of its junctions. There are sophisticated tests that can be made with oscilloscopes, curve tracers, and switching characteristics checkers, but we want to show how it is possible to test a transistor or other semiconductor device fairly completely with only an ohmmeter.

Figure 4-8. Using a Voltage to Test an Inductor for Breakdown Between Winding and Core

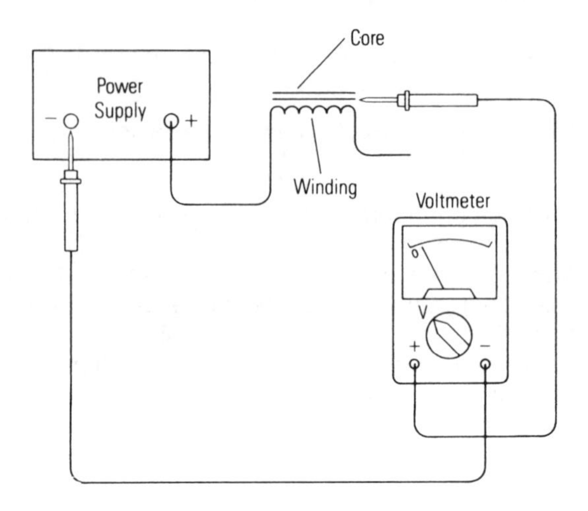

Diodes

The diode is a two-terminal, non-linear device which presents a relatively low resistance to current in one direction and a relatively high resistance in the other. A "perfect" diode would act like a switch — either on (conducting) or off (not conducting), depending on the polarity of the voltage applied to the terminals. The typical construction and circuit symbol of a diode are shown in *Figures 4-9a* and *4-9b.* The cathode of a diode is usually identified by some means of marking. On small-signal glass or plastic diodes, a colored band or dot may be used *(Figure 4-9c).* Metal can devices have a large flange on the cathode *(Figure 4-9d).* Rectifiers sometimes have a + to indicate the cathode *(Figure 4-9e).*

Figure 4-10 shows the diode being tested with an ohmmeter. Remember to identify the polarity of the multitester leads when used as an ohmmeter. Use the next range higher than R×1 to prevent excessive test current. The proper indication is a low resistance reading when the plus (+) lead is on the anode as shown in *Figure 4-10a* and a high resistance reading when the ohmmeter plus (+) lead is on the cathode as shown in *Figure 4-10b.* A low resistance in both directions indicates a shorted diode; a high resistance in both directions indicates an open diode. High current has destroyed internal connections, or high voltage has broken down the junctions.

Some DMMs, such as the one shown in *Figure 2-1,* have a special diode test function. The test is made by setting the FUNCTION/RANGE selector switch to the DIODE position. The red lead provides a positive voltage source. Connect the leads to the diode with the red (positive) to the anode and the black lead to the cathode. Read the forward voltage drop of the semiconductor PN junction in volts from the display. The diode check is performed with a maximum test current of 1.5 mA.

Figure 4-9. Diode Construction, Symbol, and Identification

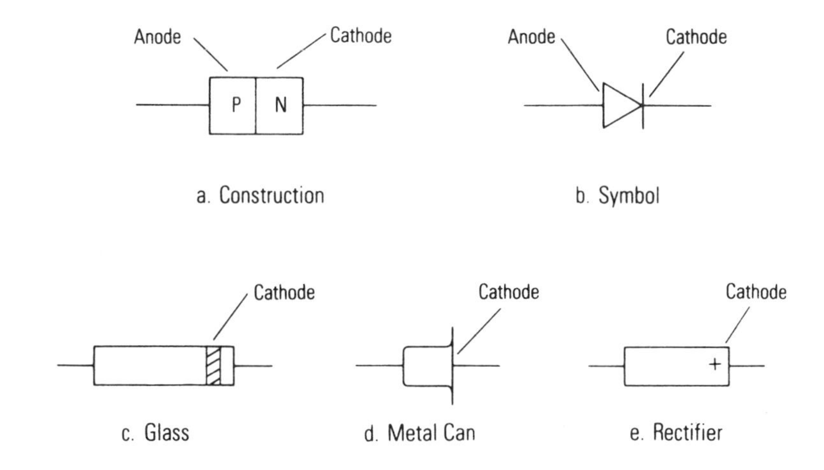

Figure 4-10. Using an Ohmmeter to Measure a Diode

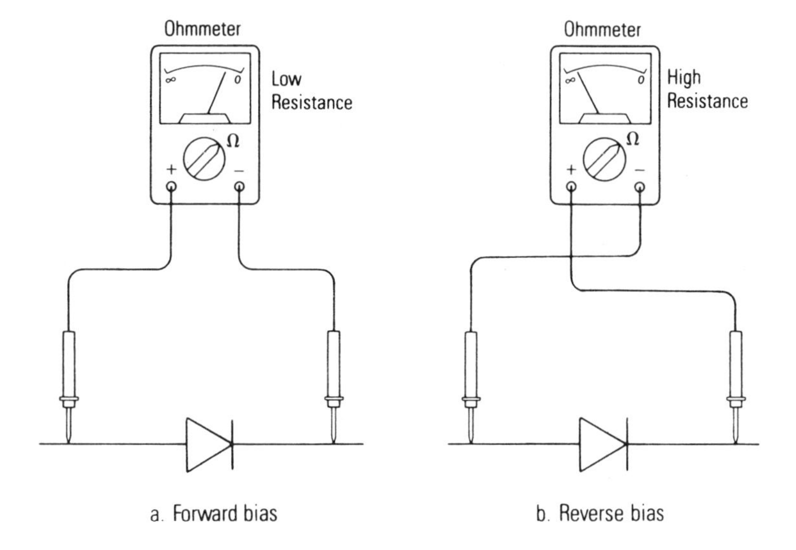

Transistors

The transistor is a three-terminal device that has virtually replaced the vacuum tube. The two basic types of transistors are the bipolar and the field-effect.

Bipolar Junction Transistors (BJT)

A transistor is a device made of two PN junctions as shown in *Figure 4-11* for both NPN and PNP transistors. The transistor is basically an OFF device and must be turned ON by applying forward bias to the base-emitter junction.

Figure 4-11. Resistance Readings Across Junctions of Transistors

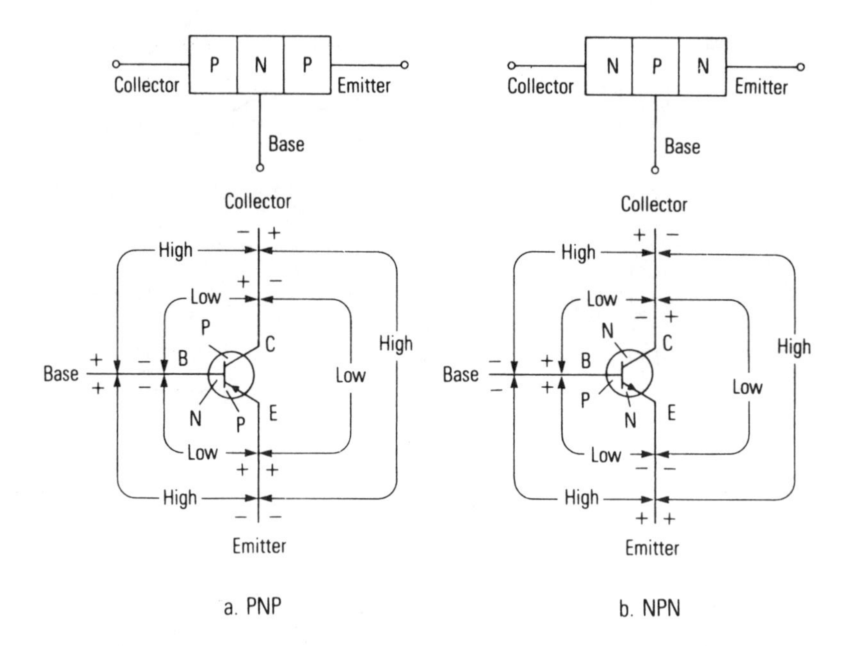

Transistors can be considered as two diodes connected back-to-back. Therefore, each junction, like a diode, should show low forward resistance and high reverse resistance. These resistances can be measured with an ohmmeter and the results should be as indicated in *Figure 4-11.* The polarities of the voltages applied are shown to indicate forward or reverse bias on the NPN and PNP transistors. *Figure 4-11* assumes the V-Ω-A(+) lead supplies a positive, and the COM (−) lead a negative voltage.

The same ohmmeter range should be used for each pair of measurements to each of the elements (base-to-collector, base-to-emitter, emitter-to-collector). For most transistors, any ohmmeter range is acceptable. However, in some meters, the R×1 range may apply excessive current for a small transistor. Also, the highest resistance range may apply excessive voltage. Either of these conditions may damage the transistor being tested. As a result, it is best to start with the middle ranges for the resistance measurements.

Defects

If the reverse resistance reading is low but not shorted, the transistor is leaky. If both forward and reverse readings are very high, the transistor is open. If the forward and reverse readings are the same or nearly equal, the transistor is defective. A typical resistance in the forward direction is 100 to 500 ohms. However, a low power transistor might show only a few ohms resistance in the forward direction, especially at the base-emitter junction. Reverse resistances are typically 20K to several hundred thousand ohms. Typically a transistor will show a ratio of at least 100:1 between the reverse resistance and forward resistance. Of course, the greater the ratio the better.

The transistor's emitter-base and collector-base junctions may be individually checked with the DIODE TEST function on a DMM like the one in *Figure 2-1.*

Operational Test of a BJT

Junction Transistor

The amplification action of a transistor may be checked with the circuit of *Figure 4-12.* It will give some indication if the transistor is operational. Normally, there will be little or no current between the emitter and collector (I_{CEO}) until the base-emitter junction is forward-biased. Therefore, a basic operational test of a transistor can be made using an ohmmeter. The R×1 range should be used. Closing S_1 allows a small base bias current to be applied from the ohmmeter internal battery through R_1. If the transistor is operational, the base current will cause collector current, thus reducing the collector-to-emitter resistance. The ohmmeter shows a decreased resistance (an increased emitter-collector current) when S1 is closed to indicate that the transistor is operational and amplifier action is taking place.

As mentioned previously, the real definitive test for a transistor is operating in a circuit. There are multitesters that measure the current gain of a transistor (h_{FE}) directly, but Chapter 5 shows the techniques to be used to measure the performance of a transistor in a circuit. Such techniques require a multitester to be used to measure voltages at various points in a circuit, and then calculating current and current gain to verify transistor performance.

Figure 4-12. Operational Test of Transistor with Ohmmeter

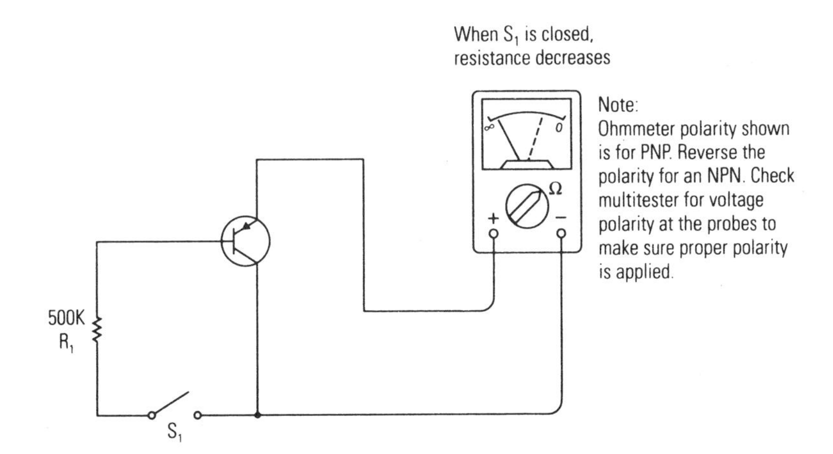

Field-Effect Transistors

A field-effect transistor (FET) is a voltage-operated device that requires virtually no input current. This gives them an extremely high input resistance. There are two major categories of field-effect transistors: junction FETs and insulated gate FETs, more commonly known as MOS (metal-oxide-semiconductor) field-effect transistors. Like the bipolar junction transistor, the FET is available in two polarities: P-channel and N-channel.

The schematic symbols for field-effect transistors are given in *Figure 4-13.* Notice the terminals are identical for N-channel and P-channel but the arrow on the gate terminal is reversed. This also indicates that the current direction from source-to-drain depends on the polarity type of the FET. Since there is no set designation for the source and drain terminals of the FET, the reference manual or the equipment schematic should be consulted to identify the terminals on the FET being tested.

Figure 4-13. Field-Effect Transistors

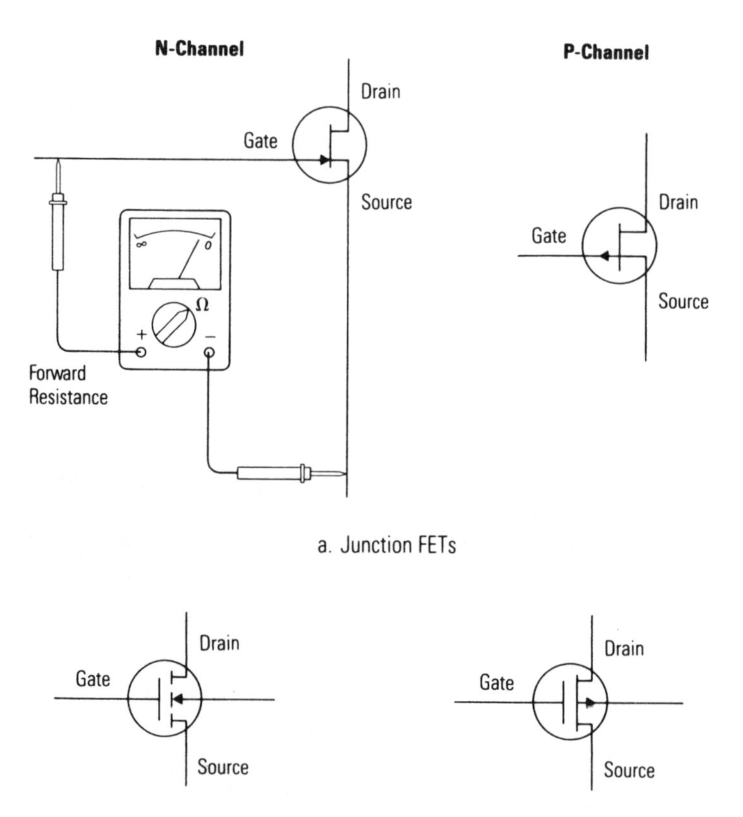

a. Junction FETs

b. MOS (Metal-Oxide-Semiconductor) FETs

Testing FETs is somewhat more complicated than a bipolar junction transistor. First, determine from all markings if the device is a JFET or a MOS type. Otherwise, the terminal measurements will have to indicate the type. Do not attempt to remove it from the circuit or handle a FET unless you are certain that it is a JFET or a MOSFET with protected inputs. If one touches the leads of these devices, static electricity can damage an unprotected device very quickly. Make certain all static electricity is grounded out before handling FETs.

JFET Measurements

To test the forward resistance of the JFET gate to source junction, use a low-voltage ohmmeter on the R×100 scale (or nearest to it). For an N-channel JFET, connect the positive lead to the gate (see *Figure 4-13a*) and the negative lead to either the source or drain. Make sure of the polarity of the ohmmeter leads. Reverse the leads for a P-channel. The resistance should be less than 1K ohms.

To test the reverse resistance of the N-channel JFET junction, reverse the ohmmeter leads and connect the negative lead of the ohmmeter to the gate and the positive lead to the source or drain. The device should show almost infinite resistance. Lower readings indicate either leakage or a short. Reverse the leads for a P-channel JFET.

Operational Test of a JFET

The following simple out-of-circuit test will demonstrate if a junction FET is operational, but will not indicate if the device is marginal. Operational means it is not shorted or open, which is by far the most common occurrence when a FET becomes defective. After the JFET has been removed from the circuit, connect the ohmmeter between the drain and source terminals. Touch the gate lead with a finger and observe the ohmmeter polarity connections to the source and drain terminals and the channel type (P or N). Reverse the leads of the ohmmeter to the terminals and again touch the gate terminal. The ohmmeter should indicate a small change in the resistance opposite to that previously observed if the FET is operational. The change in resistance will be very slight and some operational (good) FETs will not appear to change.

MOS Measurements

To test a MOSFET, the device must be handled with caution and the hands and instruments must be discharged to ground before measurements are made. If a MOSFET is to be checked for gate leakage or breakdown, a low voltage ohmmeter on its highest resistance scale should be used. The MOSFET has an extremely high input resistance and should measure "infinity" from the gate to any other terminal. Lower readings indicate a breakdown in the gate insulation. The measurements from source to drain should indicate some finite resistance. This is the distinguishing characteristic of a MOSFET; it has no forward and reverse junction resistance because the metal gate is insulated from the source and drain by silicon oxide. It should be a very high resistance with both polarities of voltage applied.

SCR Testing

An SCR is a gated diode that is used for the control of ac power. If a positive voltage is applied to the anode relative to the cathode, the diode will not conduct in the forward direction until triggered by current in the gate. Once triggered on, the diode is turned off by the voltage between anode and cathode going to zero. Testing with an ohmmeter is not recommended for high-current SCRs and should only be used as a relative indication in low-current SCRs. The current supplied by the ohmmeter may not be enough to "fire" or "hold" the SCR and, therefore, may not always indicate the true junction condition of the device.

However, a simple ohmmeter test of a low-power SCR may provide an approximate evaluation of their gate-firing capabilities by connecting an ohmmeter as shown in *Figure 4-14.* The negative lead is connected to the cathode and the positive lead to the anode (check the multitester polarity). Use the Rx1 scale on the ohmmeter. Short the gate to the anode with S1. This should turn the SCR "ON" and a reading of 10-50 ohms is normal. When S_1 is opened and the gate-to-anode short is removed, the low resistance reading should remain until the ohmmeter lead is removed from the anode or the cathode. Now, reconnecting the ohmmeter leads to the anode and cathode should show a high resistance until S_1 is closed again to short the gate to the anode.

TESTING A BATTERY

A battery is a group of cells that generate electricity from an internal chemical reaction. Its purpose is to provide a source of steady dc voltage of fixed polarity. The battery, like every source, has an internal resistance that affects its output voltage. For a good cell, the internal resistance is very low with typical values less than an ohm. As the cell deteriorates, its internal resistance increases and prevents the cell from producing its normal terminal voltage when there is load current. A dry cell loses its ability to produce an output voltage even when it is out of use and stored on a shelf. There are several reasons for

Figure 4-14. SCR Testing with Ohmmeter

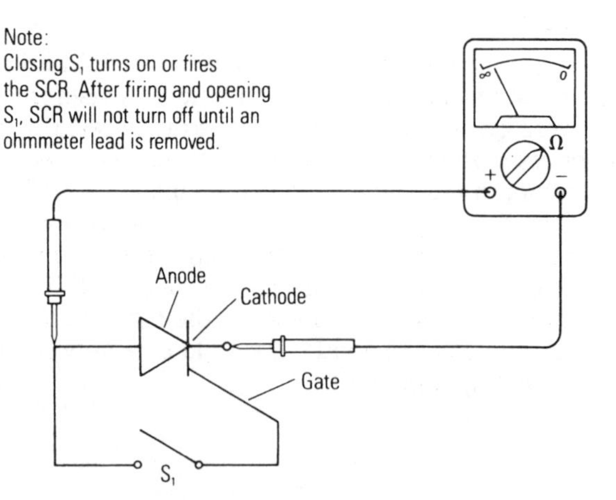

this, but mainly it is because of self-discharge within the cell and loss of moisture in the electrolyte. Therefore, batteries should be used as soon after manufacture as possible. The shelf life is shorter for smaller cells and for used cells.

Current Drain

The current drain of a battery should be measured in its actual operating condition; that is, its normal load. It can be measured by connecting the current meter in series with the battery and the equipment it is powering. The battery clip may be disconnected from the battery on one side and the current meter connected to complete the circuit. The correct polarity must be observed. Start on a high current range in case of an excessively high current due to a malfunction.

Testing Under Load

A very "weak" (high internal resistance) battery can have almost normal terminal voltage with an open circuit or no load current. Thus, a battery should be checked under its normal load condition; i.e., in the equipment that it powers with the power switch on. Out of the equipment, the only meaningful test is with a load resistor across the battery as in *Figure 4-15.* The value of the load resistor depends on the battery being tested. For a standard "D" cell, R_L = 10 ohms; a "C" cell, R_L = 20 ohms; an "AA" cell, R_L = 100 ohms; and a 9-volt battery, R_L = 330 ohms. The terminal voltage should not drop to less than 80% of its rated value under load. The internal resistance may be calculated by the equation:

$$R_I = \frac{V_{NL} - V_L}{I_L} \quad \text{where } I_L = \frac{V_L}{R_L}$$

and V_{NL} = the terminal voltage, unloaded
V_L = the terminal voltage with a load resistor
I_L = the load current

Measure V_{NL} first without a load, then measure V_L with a load, and calculate.

Figure 4-15. Testing a Battery Under Load

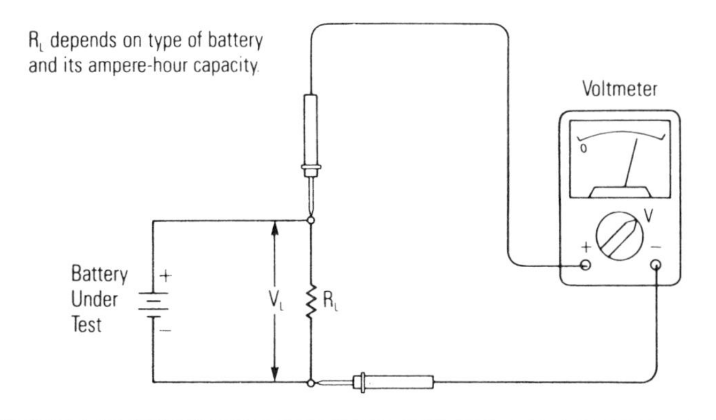

VOMs With Battery Test Function

Some VOMs have a feature that places the battery under a current load and tests its condition. The VOM shown in *Figure 4-16* has a separate range for three different classes of batteries: 125mA at 1.5V for AA and AAA cells, 250mA at 1.5V for C and D cells, and a 10mA load current test for a 9V battery. It has a Red-Green scale that shows "Bad-Good" for the cell's condition.

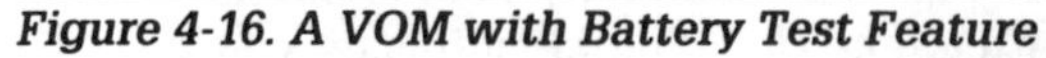

Figure 4-16. A VOM with Battery Test Feature

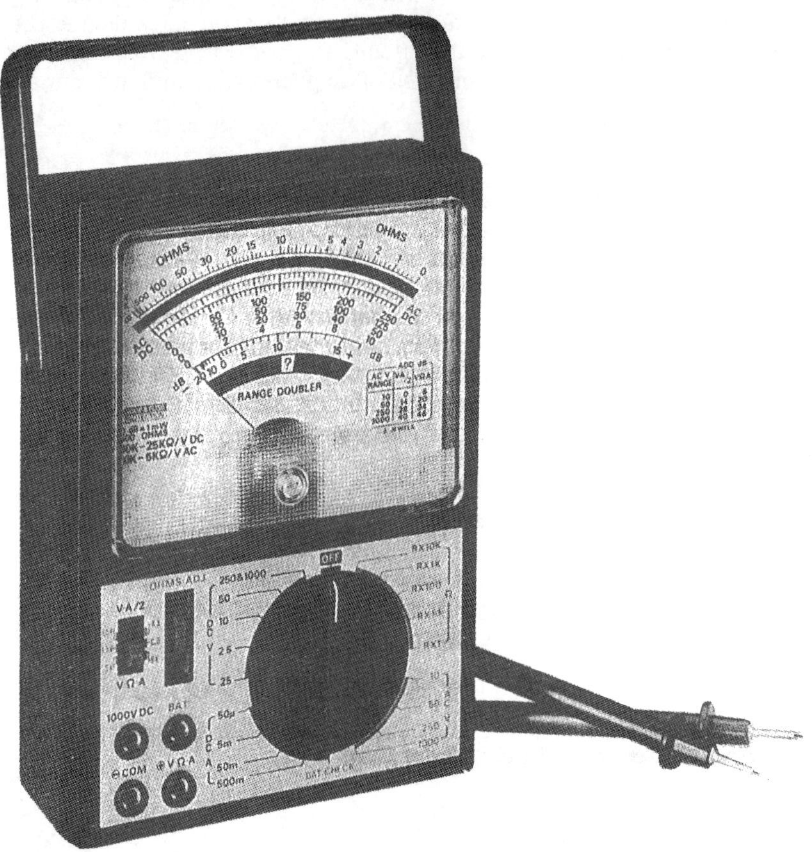

SUMMARY

Now that we have found out how to make measurements on individual components, let's look at how to measure components in circuits. That is the subject of the next chapter.

CHAPTER 5

Measuring Components In Circuits

The most popular and versatile instrument on any electronics technician's workbench or in his toolbox is the multitester or multimeter. Understanding how to use multitesters to measure individual components is important, but understanding how to use them in making measurements in a circuit, the subject of this chapter, enhances their value many times over.

ABOUT IN CIRCUIT COMPONENT TESTS

General Considerations

In-circuit measurements with a multitester usually occur with power applied to the circuit. Such measurements are commonly made because something has happened to cause a circuit, a device, or a piece of equipment to operate improperly. The measurements are made to help locate the problem. This is called troubleshooting the circuit. Because power is applied, special safety precautions should be observed. Make certain that measurement terminals are not shorted together or to ground by the test leads. This is especially so when 115VAC line voltage is being measured.

CAUTION:

Severe shock or extensive damage can occur if you are not careful.

Meter Safety

The multitester will remain a valuable servant if reasonable care is taken while operating it. Specification limits should not be exceeded. When the value to be measured is unknown, begin with the multitester set on the highest range; then move to lower ranges until the meter has an indication which is near full-scale.

CHECKING A POWER SUPPLY

Each electronic system has a power supply, even if it is simply a battery. Let's look at a typical simple power supply and examine how its voltages can be measured.

Basic Circuit Understanding

Typically, electronic power supplies consist of rectifiers, diodes, transformers, a filter made of capacitors and inductors or resistors, and a bleeder resistor, which can be used also as a voltage divider. *Figure 5-1* shows a popular power supply circuit. This circuit uses a full-wave, center-tapped rectifier circuit to convert the ac voltage delivered by the transformer to dc. The pulsating current from the rectifier is smoothed by the filter. The bleeder resistors serve at least

Figure 5-1. A Popular Power Supply Circuit

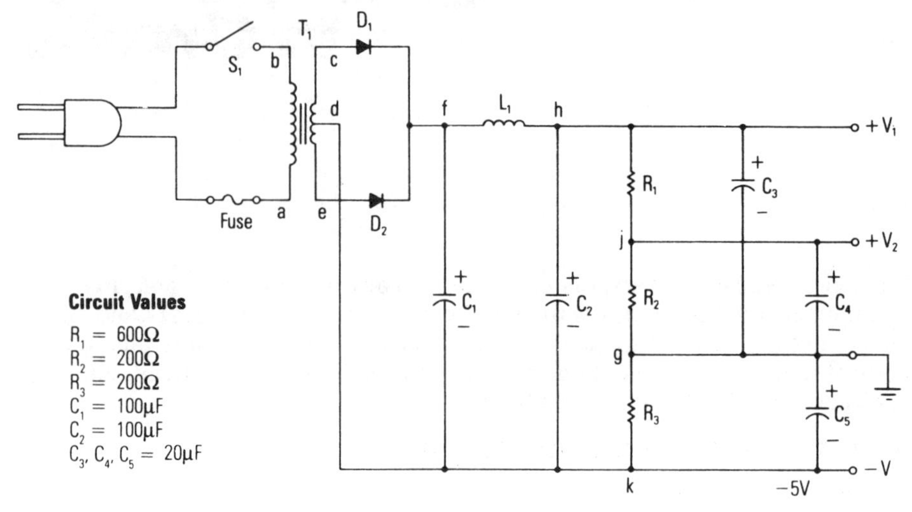

two purposes here. They provide multiple outputs, including one negative voltage with respect to ground. They also provide a constant minimum load for the power supply to reduce output voltage variations.

Normal Operation Reference Values

Normal values of ac voltage, dc voltages, and resistances to ground expected in the circuit are very valuable reference values to a technician troubleshooting a power supply. *Table 5-1* shows some typical values for the circuit of *Figure 5-1* that should prove very useful.

Table 5-1. Normal Voltage and Resistance Measurements

VOLTAGE MEASUREMENTS			RESISTANCE
Type	Voltage Test Points	No Load Voltage Value (V)	Resistance a thru k to g
ac	a-b	110	∞
ac	c-e	40	*$R_3 \parallel R_1 + R_2 = 160\Omega$
ac	c-d	20	*$R_3 \parallel R_1 + R_2 = 160\Omega$
dc	f-g	+20	$R_1 + R_2 = 800\Omega$
dc	h-g	+20	$R_1 + R_2 = 800\Omega$
dc	j-g	+5	$R_2 = 200\Omega$
dc	k-g	−5	$R_3 = 200\Omega$

*Diodes forward biased

Locating Defective Components

In the circuit of *Figure 5-1*, suppose that there is no voltage between point h and ground. To isolate the problem, the ac voltage is measured between points c and e to determine if the correct input ac voltage is present.

If not, the trouble could be due to any of the following:

a. no power at ac outlet
b. open line cord or line plug
c. open fuse

d. defective switch (remains open when it should be closed)
e. open transformer winding

Voltage, continuity and/or resistance measurements in the above areas should isolate the problem to one component.

If the correct ac voltage is present at points c and e, then it is likely that the defect would be in:

a. the load
b. the rectifiers
c. the filter inductor
d. an open in the wiring between points c-e and h

Disconnect the load from points V_1, V_2, and $-V$. If the voltage at h returns, then the problem is in the load. If not, check b (the rectifiers), c (the inductor) or d (open wiring). Voltage and resistance measurements to ground with *Table 5-1* as a reference should isolate the problem.

Resistance Measurements

CAUTION:
Before measuring resistance, turn off power and discharge all capacitors, especially if they are electrolytic.

Specific circuit test points are selected so that resistance measurements can be made from the test points to ground to determine if there are circuit shorts or circuit opens. For example, if one of the filter capacitors is shorted (say C_3), the resistance from point h to ground will be zero. When the trouble is found and the faulty component located and replaced, perform an operational check on the supply to make sure it is completely repaired.

TROUBLESHOOTING A SIMPLE AC CIRCUIT

Doorbell Circuit

Figure 5-2 shows a typical doorbell circuit. This circuit has the doorbell and switch in series with the transformer secondary winding. The doorbell operates at 10VAC. This low voltage is not dangerous, and as long as the measurements are made on the secondary side of the transformer, it is not necessary to turn off the circuit breaker to check out the circuit. The voltage and resistance measurements for the normally working doorbell circuit of *Figure 5-2* are shown when the button is not pushed and when the button is pushed. M_1, M_2 and M_3 are different meter readings. If any of the voltages are not correct, the listed readings should help to determine the circuit problem. If not, then resistance measurements will have to be made. *CAUTION: Turn off the circuit breaker (or remove fuse) before making these resistance measurements.*

When The Bell Doesn't Ring

Assume that the bell does not operate when the button is pushed. The two most common problems are a bad switch or a bad bell. To track down the trouble, measure the voltage across the push button switch. With the switch open, 10 volts should appear across the open switch (M_1). There is no current in the circuit, thus, no voltage across the bell (M_3). When the switch is closed (button pushed), M_1 should read zero. If a voltage appears across the switch when the button is pushed, the switch is probably defective. Check the

Figure 5-2. A Typical Doorbell Circuit

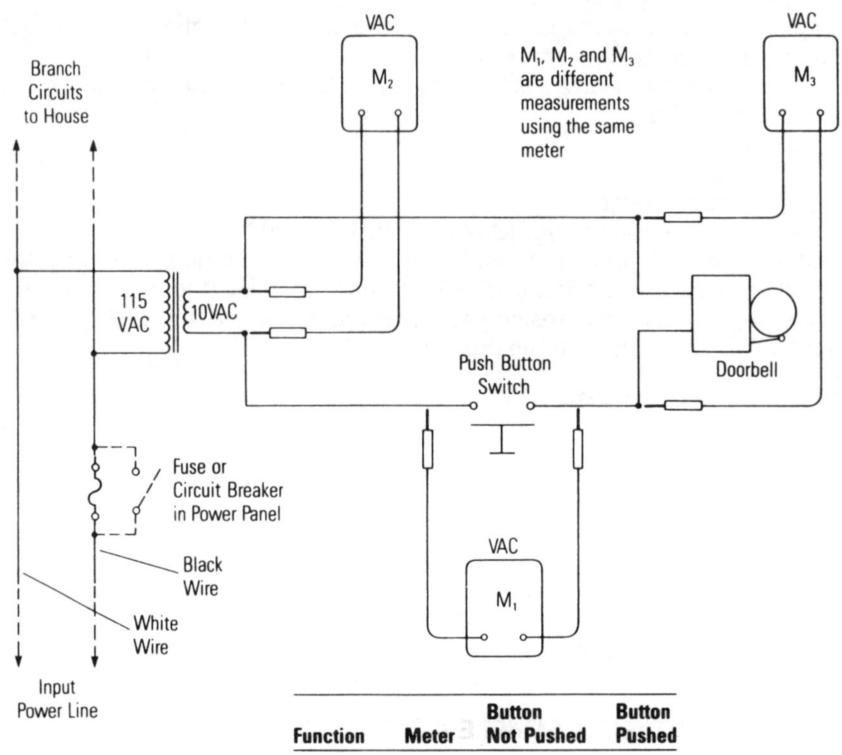

Function	Meter	Button Not Pushed	Button Pushed
Voltage	M_1	10V	0V
	M_2	10V	10V
	M_3	0V	10V
Resistance	M_1	4-7Ω	0Ω
	M_2	1-2Ω	0-1Ω
	M_3	2-5Ω	0-1Ω

contacts to see if they are corroded or broken. The switch can possibly be repaired by simply scraping and cleaning the contacts; however, it may have to be replaced. If 10 volts appears across the bell when the button is pushed and the bell does not ring, the bell is probably defective. Disconnect it and check its resistance to see if it has an open coil.

Tracing a Short Circuit

Now let's assume that instead of an open in the circuit, there is a short in the circuit. We'll use voltage readings to find the short in the circuit. Assume that M_1 reads the normal zero volts when the button is pushed. However, M_1 reads an abnormal value of only 2-3 volts when the button is not pushed. It should read 10 volts. Now, measure the voltage across the transformer secondary (M_2). It also is a low reading of 2-3 volts. You note that the transformer is quite warm. You suspect a short, so you disconnect one of the wires at the secondary terminals, and the voltage across the secondary (M_2) jumps to 10 volts. This indicates that there is a short circuit somewhere in the wires going to the doorbell or in the doorbell itself. This short circuit drains so much current from the transformer secondary that it is overloaded and the output voltage is reduced to the low value of 2-3 volts.

To locate the trouble, leave the wire disconnected at the transformer and disconnect one of the wires at the doorbell terminals. Measure the resistance across the two wire ends at the doorbell end. (You do not need to turn off power to make these resistance measurements.) If it is not infinite, the short is in the wires. If possible, examine the wires for frayed or worn insulation and possible points where they may touch. Sometimes a nail is driven through the wires, scoring the insulation and causing the wires to short out with age. If you can find the short by visual inspection, clear it and check for normal voltages as indicated above. It may be simpler and faster just to replace the entire wire run.

If the resistance is infinite across the wire ends, then verify by resistance measurement that the short is in the doorbell and replace the doorbell.

TROUBLESHOOTING YOUR TELEPHONE INSTALLATION

The fact that we can pick up our telephone and talk to almost anyone in the world is surely a modern miracle. Its dependability is amazing; however, there can be problems. Though you are not allowed to do repair work on the equipment owned by the local telephone company, you can make simple tests on the line and your own telephone equipment.

Testing the Telephone Line

A good way to determine if the trouble is on the telephone line or in one of the telephone sets is by substitution. If you connect a known good telephone to the line and it works, you know that the first telephone set has a problem and the line is OK. But if the substitute does not work, then you have verified that the line has a problem. If the trouble is on the line from the telephone central office to where it enters the house, the telephone company is responsible for the repair and will do it without charge. If the trouble is in your house, you can repair it yourself. (If you don't do it yourself, and if you don't have a service contract with the telephone company, then you will have to pay for the repair.)

On-Hook and Off-Hook

The first connection inside the house is usually made at a type 42A terminal block as shown in *Figure 5-3a*. *Figure 5-3b* shows how five phone jacks are wired from the block: one directly and four on branch (extension) lines. The incoming line and the branch lines can be tested with the voltmeter of a DMM. The voltage between the red and green wires should be about 48 volts dc at the entry block and at each jack with all of the telephones either on-hook (hung up) or unplugged from their jacks. Connect the minus (−) meter lead to the red wire (ring) terminal and the plus (+) meter lead to the green wire (tip) terminal. An open or a short on the line can result in a zero voltage reading at a particular jack.

The voltage measured across the line in the off-hook condition (handset or "receiver" lifted and ready for use) will depend on several things, but mainly the distance from the central office and the wire size used for the phone line. Typically, the voltage will be from 5 to 10 volts dc.

Figure 5-3. Initial Connection of Incoming Telephone Line

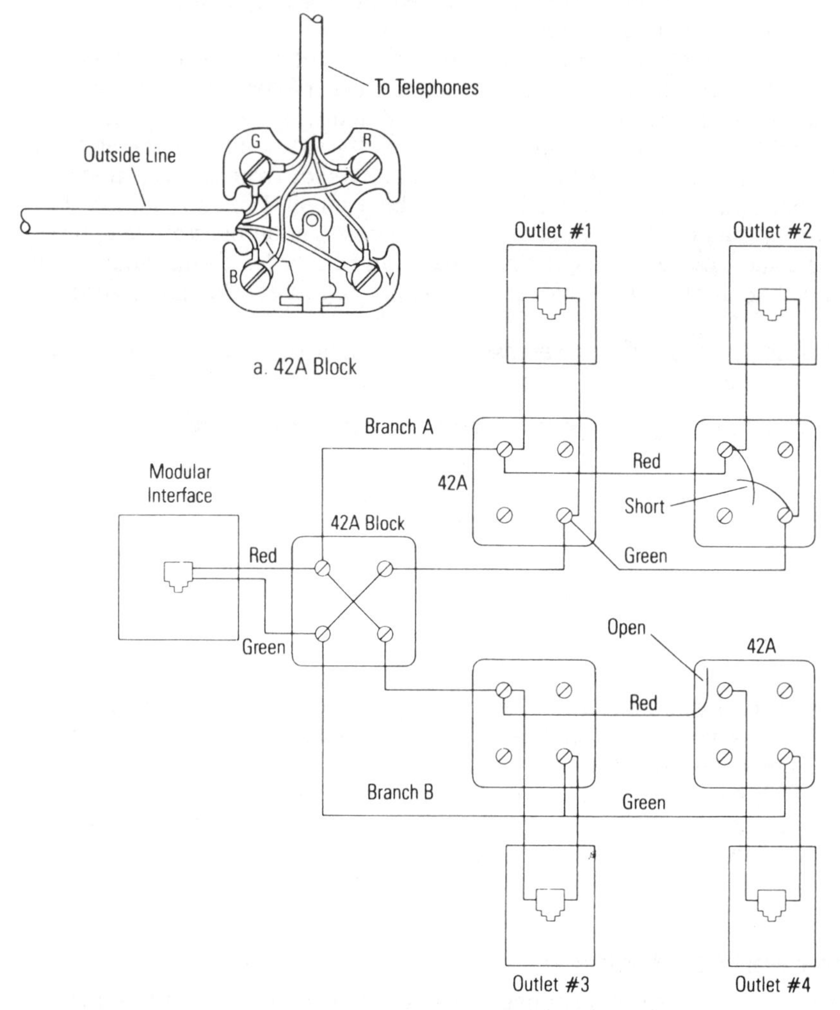

Testing Branch Circuits

Let us now consider how to find a short or open at some point on one of the lines as shown in *Figure 5-3b.*

There are two legs that form branches from the 42A block: Branch A has outlets #1 and #2 and Branch B has outlets #3 and #4. Let us assume that if a modular telephone is plugged into any of these four outlets, it will not work. First, we will determine that the trouble is in the house and not on the phone line from the central office.

At the 42A terminal block where the line enters the house, disconnect all the green branch wires and measure the voltage. (The green wires are chosen arbitrarily: you could just as well disconnect only the red wires.) As mentioned previously, it should be about 48 volts with the red wire negative (−) and the green wire positive (+). Let us assume this is as it should be. Connect a known good telephone set across the incoming line. Getting a dial tone when the receiver is off-hook confirms that the incoming line is OK. So the problem must be in the house wiring.

Reconnect the green wire for Branch A (leave Branch B disconnected) to the 42A and the voltage drops to zero (or very near zero). This means a short somewhere on Branch A. Go to outlet #1 and visually inspect the connections. You do not see any short. Disconnect the green wire that feeds outlet #2 and the voltage at outlet #1 is 48 volts. Connect a modular telephone to outlet #1 and it works fine. Reconnect the feed line to outlet #2 and again the voltage drops to zero. Go to outlet #2 and visually inspect the connection. You see that the excessively long stripped ends of the red and green wires are touching and creating the short circuit. Pull the wires apart and cut off the excess length to clear the short. Connect telephones to outlet #1 and outlet #2 and verify that both work OK.

Now reconnect the green wire for Branch B. The voltage at the entry block is still 48 volts. Go to outlet #3 and measure the voltage. It is 48 volts. Connect a telephone to outlet #3 and it works OK. Go to outlet #4 and measure the voltage. It is zero volts. Since you measured 48 volts at outlet #3 even with the feed wires to outlet #4 connected, you suspect an open somewhere between outlet #3 terminals and outlet #4 terminals. Although the wires at outlet #4 look OK, you try moving them around a little. You find that the red wire has broken right at the terminal but it has remained in position so it looked OK at first. Fortunately, the red wire has some slack so you can just strip the insulation and reconnect the wire. The voltage is now 48 volts and when you connect a telephone to the jack, it works OK. Now the telephones at all jacks are working properly.

Testing the Telephone Set

If it was determined that a telephone set is not operating properly, three problems are very common. A switch contact is bad, the talk and listen circuits are faulty, or the ringing circuit does not operate. Three quick tests can be made to isolate these problems.

Measuring Resistance

To determine if the switchhook contacts are operating properly, check the resistance of the telephone set with an ohmmeter. Disconnect the telephone set from the line and measure between the red and green wires of the telephone set as shown in *Figure 5-4*. With the handset on hook, the resistance should be infinity; with the handset off hook, the resistance should be 2,000 to 10,000 ohms. Any reading below 1,000 ohms indicates some problem with the telephone set. Further detailed troubleshooting would be necessary to locate the problem.

With the handset off hook, repeatedly touch and remove the ohmmeter leads to the red and green wires. You will hear a clicking sound from the receiver of the handset if the talk and listen circuits are working properly. If the transmitter is an electrodynamic microphone, the clicks also will be heard from the mouthpiece.

Figure 5-4. Measuring the Handset with an Ohmmeter

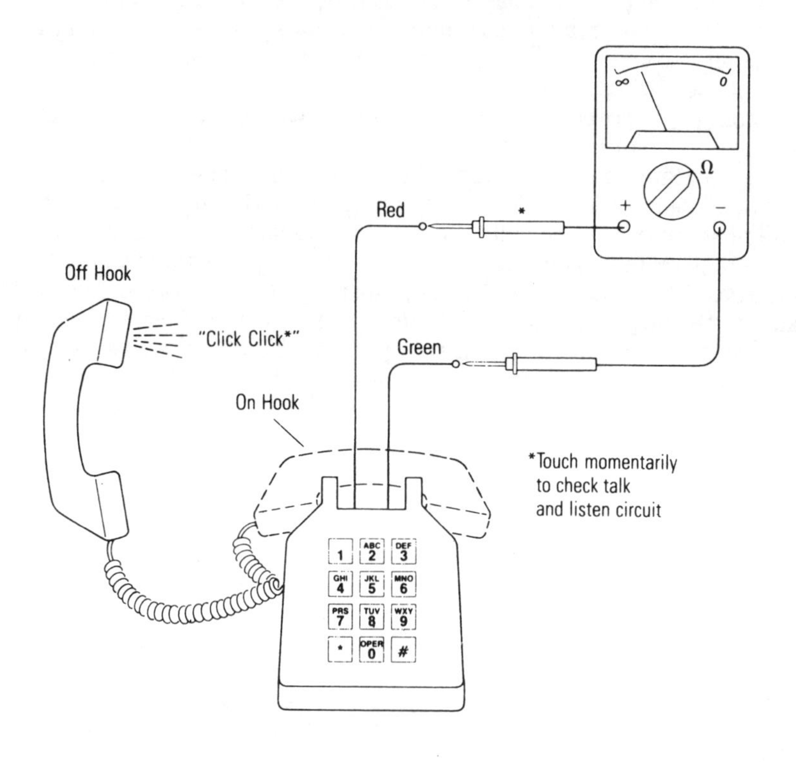

Figure 5-5. Measuring Ringing Voltage

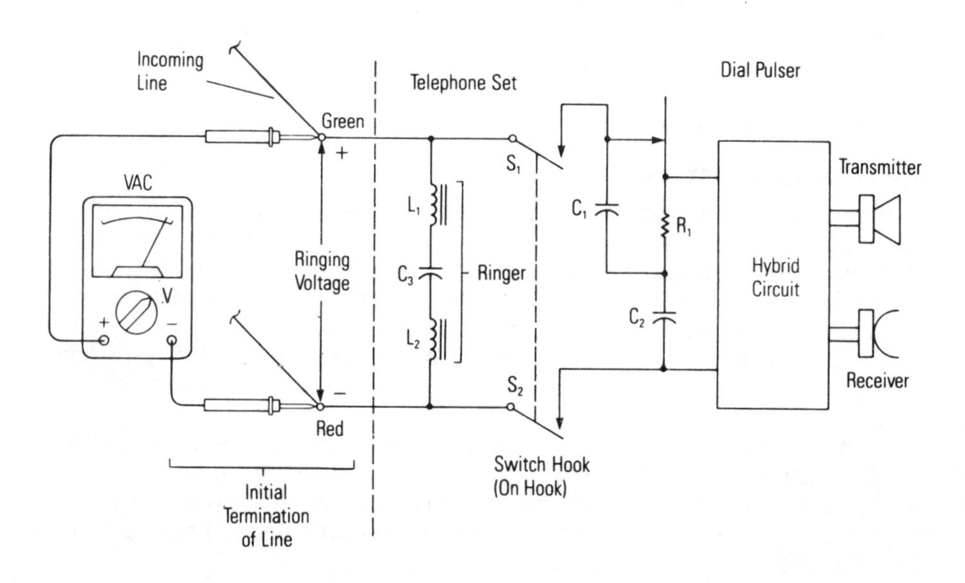

Test of the Ringer Circuit

The easiest way to test the telephone set for ringer problems is to see if the telephone rings when ringing voltage is available. With the telephone set connected to the line, the handset on hook, and a voltmeter connected across the line as shown in *Figure 5-5* at the 42A block (or the initial entry point into the house or apartment), have someone call your number. The central office applies a ringing voltage of about 90 volts ac across the line to ring the telephone. The voltage at the telephone set depends on the distance from the central office. *Be careful not to touch the wires with your hands or fingers. The ac ringing voltage can shock you.* If the meter indicates that ringing voltage is present but the telephone does not ring, then the ringer circuit in the telephone set is defective. More detailed troubleshooting would have to be done to locate the specific problem with the ringer.

OPERATION OF A BJT CIRCUIT

Normal Operation

Figure 5-6 is a typical kind of circuit that uses a bipolar junction transistor (BJT), in this case a NPN transistor (see *Figure 4-14* for symbol), to be the active device in an amplifier circuit. Certain conditions must be met to make the transistor circuit operate properly. The base-emitter junction must be forward-biased in order for the transistor to be turned on and conduct current. If the base-emitter junction is reverse biased (lack of forward bias) the transistor will be turned off and will not conduct current.

The letters NPN or PNP tell us much of what we need to know about the voltages in the circuit for normal operation. The P material must be positive

Figure 5-6. Transistor Amplifier for Normal Operation

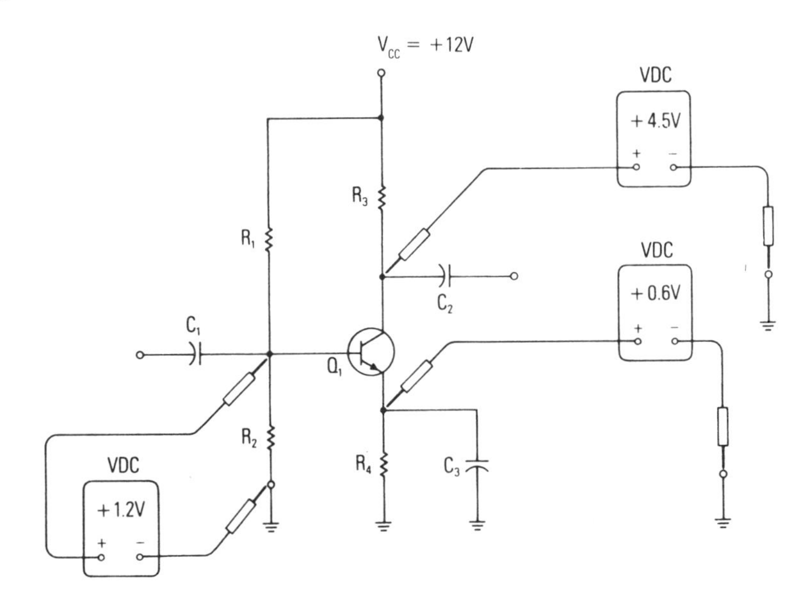

and the N material must be negative to forward bias a PN junction. The center letter, N in PNP and P in NPN, indicates the base of the BJT transistor. Thus, for a NPN transistor to have a forward-biased base-emitter junction, the base (P) must be positive with respect to the emitter (N).

The letters also tell us the polarity of the power supply applied to the collector in a BJT amplifier circuit, since the collector-to-base junction must be reverse biased. If the transistor is a NPN, a positive voltage is applied to the collector with respect to the base and vice versa for a PNP. As the base voltage is increased with respect to the emitter, the base current increases, which causes the collector current to increase. Conversely, as the base voltage is decreased with respect to the emitter, the base and collector currents decrease. When the base-emitter voltage decreases to less than the cutoff voltage for the transistor, the currents stop completely.

A conducting transistor normally has a relatively high resistance between collector and emitter until it is driven by increased base current into the saturation region (the region where the collector is only a few tenths of a volt away from the base). In a saturated transistor, the emitter-collector resistance is very low — almost a short circuit — and further increase in base current (forward bias) does not cause any further increase in collector current. Conversely, a transistor that is cutoff; that is, the base-emitter junction is not forward biased, has a very high resistance — almost an open circuit — and only a very small leakage current exists.

Measuring Voltages

An easy way to evaluate whether a transistor is operating properly in a circuit such as *Figure 5-6* is to measure the voltages at the collector, base and emitter while the transistor is in the circuit. *Figure 5-6* gives typical voltage values when measuring the voltage from ground to the respective transistor elements of a NPN BJT.

The voltages could have been measured between the transistor electrodes, that is, from element to element, but the most common way, and the way most voltage values are given in service manuals and on schematics, is to use ground as a common reference as shown in *Figure 5-6*. Ground is often, but not always, connected to the metal chassis into which the circuit is mounted. Also, one terminal of the power supply is often connected to ground. This is implied in *Figure 5-6* because the V_{CC} is shown as +12 volts with no specific reference given for the negative terminal.

Junction Voltages

A forward-biased **silicon** transistor normally has a difference of 0.6 to 0.8 volt between the emitter and base. The polarity depends on whether the transistor is NPN or PNP. As mentioned previously, the polarity will be in the direction to forward-bias the base-emitter junction (positive on P material and negative on N material) and to reverse-bias the collector-base junction (positive on N material and negative on P material).

A forward-biased **germanium** transistor normally has a difference of 0.2 to 0.3 volt between the emitter and base. The types and polarities are the same as for silicon.

Steady-State No-Signal Voltages

Voltages given in *Figure 5-6* are the steady-state no-signal bias voltages. The transistor amplifier will usually be operated Class A, which means that the input signal will cause the base and collector currents to increase and decrease linearly around a fixed no-signal bias level — the level due to the no-signal bias voltages. Little change will occur in the emitter-to-base voltage, but large changes can occur in the collector-to-base (or collector-to-ground) voltage due to the voltage drop across the load resistor, R_3, caused by the signal current change. The amount of current change is an indication of the amplifier's gain.

Parallel Resistance Check

One of the simplest tests to determine that the transistor is operating properly in the circuit is to place a resistor R_X equal in value to R_1 across R_1 (in parallel) in the circuit as shown in *Figure 5-7.* The voltmeters in *Figure 5-7* show increased base-to-ground voltage, increased emitter-to-ground voltage, and decreased collector-to-ground voltage. These changes in voltage indicate that the amplifier is operating. The increased base-to-ground voltage, which causes increased base-emitter current, is due to the effective reduction of R_1. The collector-to-ground voltage is decreased due to the increased voltage drop across R_3 which is due to the increased collector current resulting from the increased base current. The emitter-to-ground voltage is increased due to the increased base and collector current through R_4.

Figure 5-7. Increasing Transistor Collector Current by Increasing Base Current

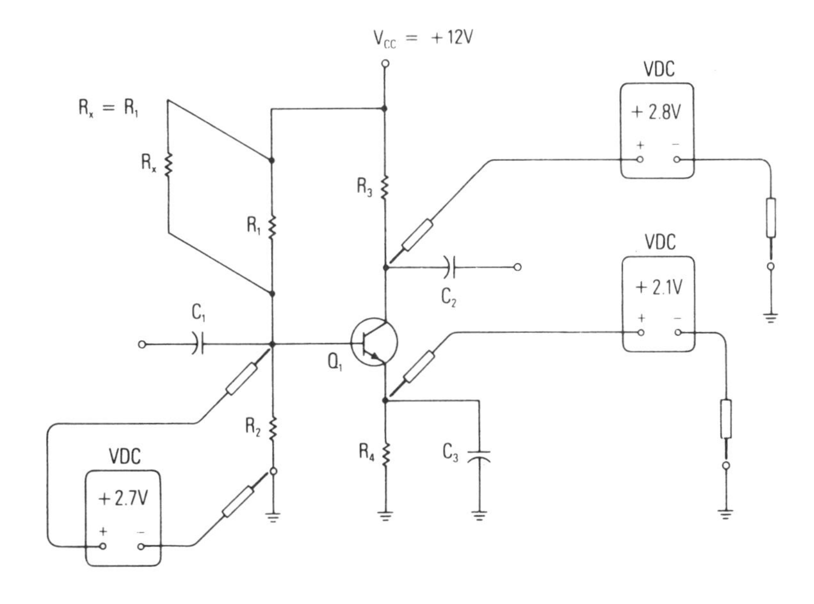

Measuring Current Gain

Transistors basically are current amplifying devices. The small-signal current gain is the gain that is active in producing the amplification in a Class A small-signal amplifier like that shown in *Figure 5-6*. Even though the small-signal current gain and the large-signal current gain, h_{FE}, are not directly correlated, the h_{FE} current gain can be used as an indicator of the relative small-signal current gain available from a transistor. To determine the h_{FE} current gain, measure the base current, I_B, and collector current, I_C, as shown in *Figure 5-8*. Calculate the h_{FE} current gain, commonly called "Beta," by dividing I_C by I_B.

The common problem with the circuit in *Figure 5-8* is that circuit leads must be broken to insert the current meters. To avoid breaking the circuit, voltage measurements can be made across resistors, as shown in *Figure 5-9*, and by using Ohm's law, the current through the resistors can be calculated. For example,

$$I_1 = V_1 / R_1$$
$$I_2 = V_2 / R_2$$
$$I_3 = V_3 / R_3$$
$$I_4 = V_4 / R_4$$

The collector current is I_3, the emitter current is I_4, and the base current, I_B, is $(I_1 - I_2)$

Therefore, calculate h_{FE} or Beta by:

$$h_{FE} = I_C / I_B = I_3 / (I_1 - I_2)$$

DEFECTIVE TRANSISTORS

The most common causes of defective transistors are: (1) the transistor junction opens because of excessive current which causes overheating, and (2) the transistor junction shorts because of a breakdown caused by excessive voltage.

Figure 5-8. Measuring Collector Current and Base Current to Calculate Current Gain

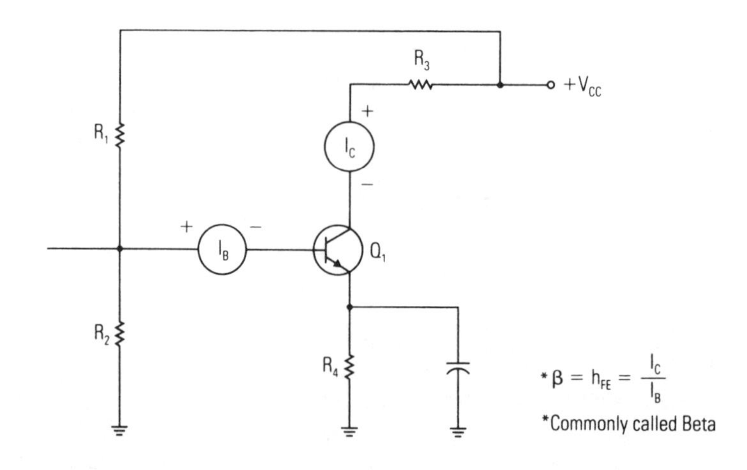

Figure 5-9. Using Voltage Measurements to Calculate Currents

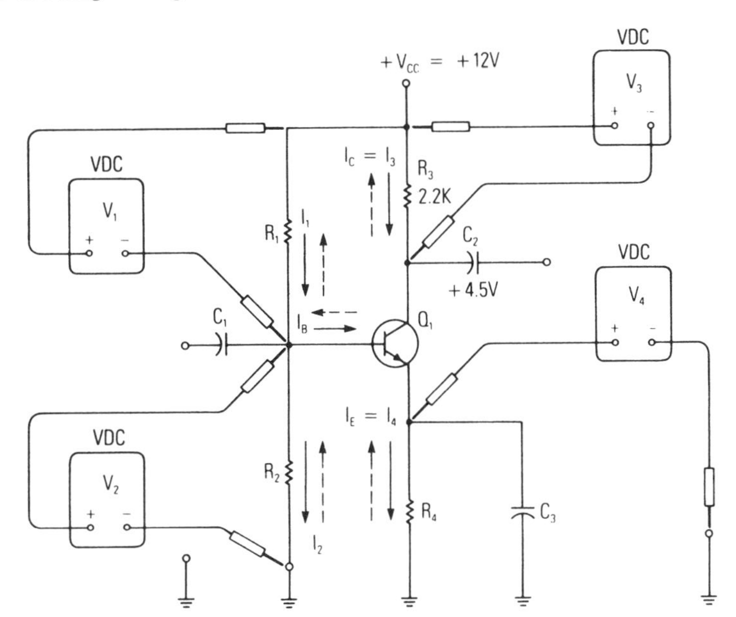

Open Junctions

If a junction opens, the transistor will no longer draw current. The most obvious and quickest way to detect this condition is to measure the collector voltage. If the collector does not draw current, the collector voltage will be very near to the supply voltage. This voltage will be considerably higher than the voltage shown in *Figure 5-6* as the normal no-signal steady-state operating voltage. There may not be much difference in the base voltage if the base-emitter junction opens because its voltage is commonly set by the bias resistor network of R_1 and R_2. Because the bias current is much greater than the base-emitter current, stopping the base-emitter current has little effect on the base voltage.

Shorted Junctions

If a junction is shorted, more than normal current is likely to flow and operating voltages will be affected accordingly. Suppose the collector-to-base junction is shorted. Now the collector voltage is likely to be much lower than normal because the current is controlled by the short circuit and not by the transistor. In addition, the base voltage is likely to be much different because of the loss of isolation of the collector junction.

If the emitter-to-base junction is shorted, the transistor action will stop. As a result, the collector voltage should be very close to the supply voltage, the emitter voltage will be the same as the base voltage, and the transistor is disabled. This case is shown in *Figure 5-10*, where a physical short has been placed across the base and emitter with a clip lead. The voltage measurements are shown.

Figure 5-10. Shorting Base-to-Emitter to Cut Off the Transistor

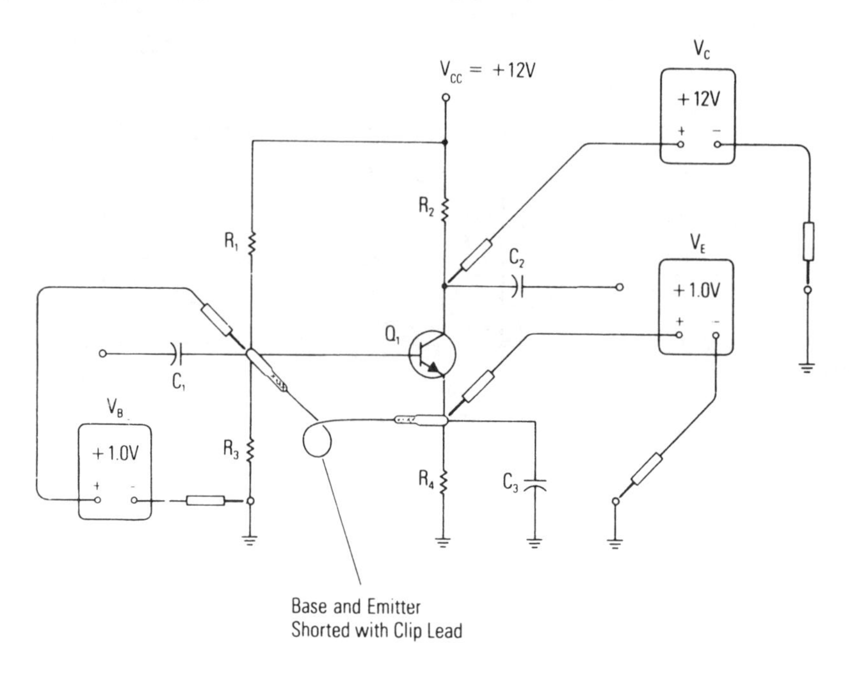

DEFECTIVE RESISTORS

Open R_1

Even though some of the cases are rare, let's demonstrate how BJT circuits can be analyzed considering the effect of certain component failures; that is, in particular, an open resistor or a shorted or open capacitor. What would happen if R_1 of *Figure 5-6* were to open?

The results are shown in *Figure 5-11.* If a resistor doesn't have current through it, there will not be a voltage drop across it. With R_1 open, there is no voltage drop across R_2. Thus, the no-signal steady-state condition of the transistor is: the base-emitter is cut off (not forward-biased) and the collector is at the supply voltage. Any negative-going portion of the input signal would be clipped off. Some positive-going signal might be coupled through the transistor and appear as a grossly distorted signal at the collector.

Shorted C_3

A shorted C_3 is equivalent to a shorted R_4 since they are in parallel. Because capacitor shorts are common (resistor shorts are rare) this condition is more likely to occur than even an open resistor. We represent this case by shorting across C_3 with a clip lead as shown in *Figure 5-12.*

The emitter voltage is zero. The transistor, heavily forward-biased, saturates and passes a current limited by V_{CC} divided by R_3. This normally prevents the transistor from being damaged. The base voltage is clamped 0.7 volt above the emitter. The collector voltage is 0.1 volt above the emitter because this is the approximate saturation voltage of a silicon transistor.

Figure 5-11. Voltages as a Result of an Open R_1

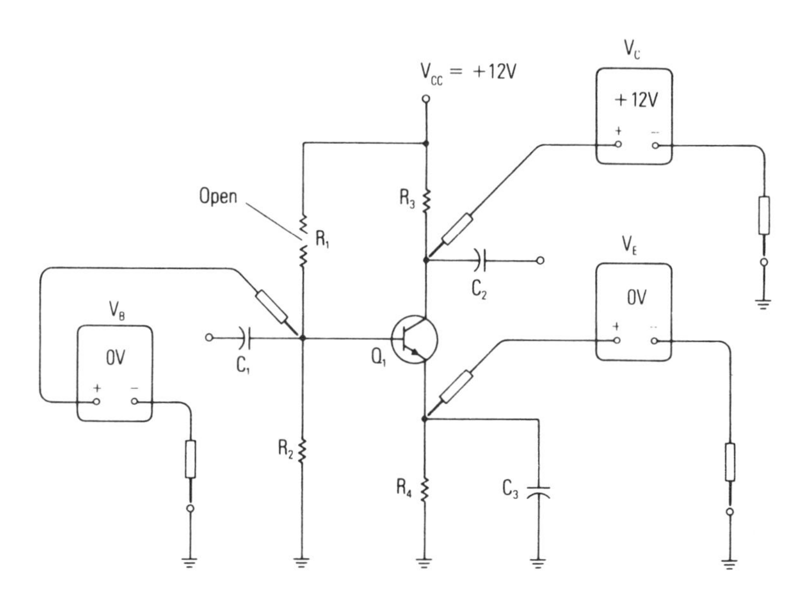

Figure 5-12. Voltages Resulting from a Shorted C_3

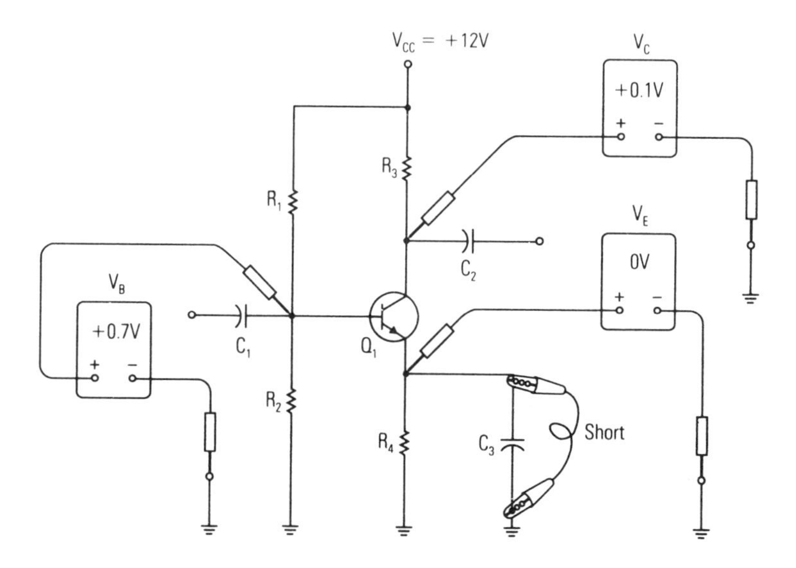

Open C_3

C_3 provides an ac signal bypass around the bias stabilization resistor R_4. If C_3 is open, an ac signal voltage will appear across R_4, introducing negative feedback into the amplifier circuit. The gain of the stage will be reduced by a value close to the ratio of R_3/R_4. The steady-state no-signal dc voltages will not change. A quick check is to connect a known good capacitor of the same value in parallel with the C_3.

CAUTION:

Turn off the power before connecting the parallel capacitor to prevent a large surge current from damaging Q_1. Then reapply power.

A FIELD-EFFECT TRANSISTOR AMPLIFIER

A normal operating FET amplifier is shown in *Figure 5-13.* The normal operating no-signal steady-state voltages are shown. Although the functions of the capacitors in the FET amplifier circuit are similar to those in the BJT amplifier circuit, the results of failure are not necessarily the same. The values of capacitance are much smaller due to the higher impedances in the FET circuit. If C_1 were to short in the FET circuit of *Figure 5-13,* the voltage from the preceding stage would be coupled to the gate and would increase the current through the FET.

In some cases, C_1 will not be shorted but will be leaky; that is, it will act as a high resistance to dc and allow some current to pass. Any voltmeter reading above zero at the gate indicates a leaky C_1. With a leaky C_1, only a

Figure 5-13. A Typical FET Amplifier Circuit

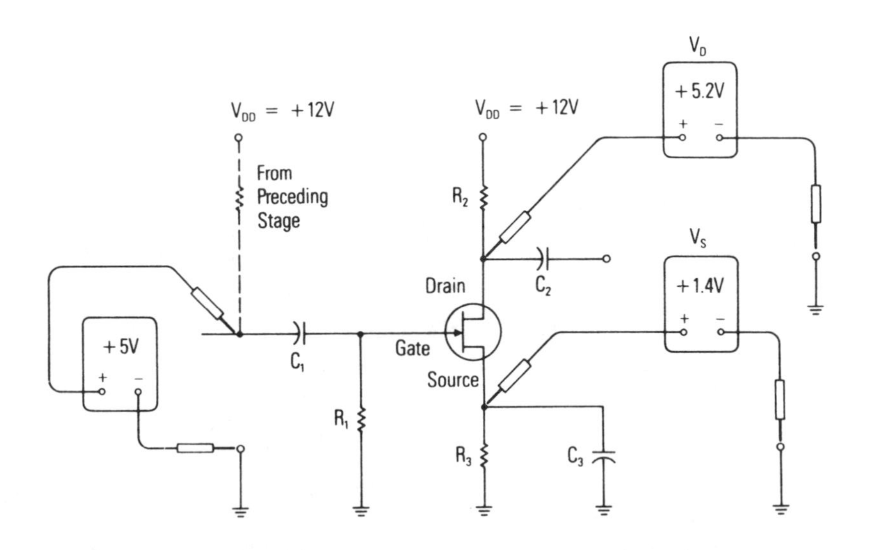

small portion of the voltage coupled from the previous stage will appear on the gate, but the effect is the same. A leaky C_1 reduces the reverse bias on the gate, causing the drain current to increase. This will cause the drain voltage to be lower than the steady-state value and restrict the signal swing.

TYPICAL AMPLIFIER IN RADIO OR TV

Figure 5-14 shows a transistor amplifier circuit that might be designed into a television or radio receiver. Note that although the power supply voltages within the system are quite a bit higher than in the previous circuits, resistor divider circuits are used to reduce the supply voltage for the circuit to the normal static voltages used in the previous circuits. R_8 may or may not be present.

GROUND REFERENCE AND CIRCUIT CONFIGURATION

Sometimes circuits are wired or drawn so that they look different from *Figure 5-6*. However, as far as the transistor is concerned, the operating conditions are the same. For example, compare *Figure 5-15* with *Figure 5-6*.

In *Figure 5-6*, the negative power supply terminal is connected to ground. In *Figure 5-15*, the positive power supply terminal is connected to ground. Also, the top base-bias resistor, R_1, and collector load resistor, R_3, are connected to ground. The emitter resistor, R_4, and the bottom base-bias resistor, R_2, are connected to the negative supply terminal. It is important for you to realize that **the bias voltages between the transistor elements are identical** in the two figures — and that is what matters. If ground is used as a reference (as it is in *Figure 5-15*), the measured voltages will be negative. If the negative supply terminal is used as a reference, the measured voltages will be positive.

Figure 5-14. A Typical Amplifier in a TV or Radio Set

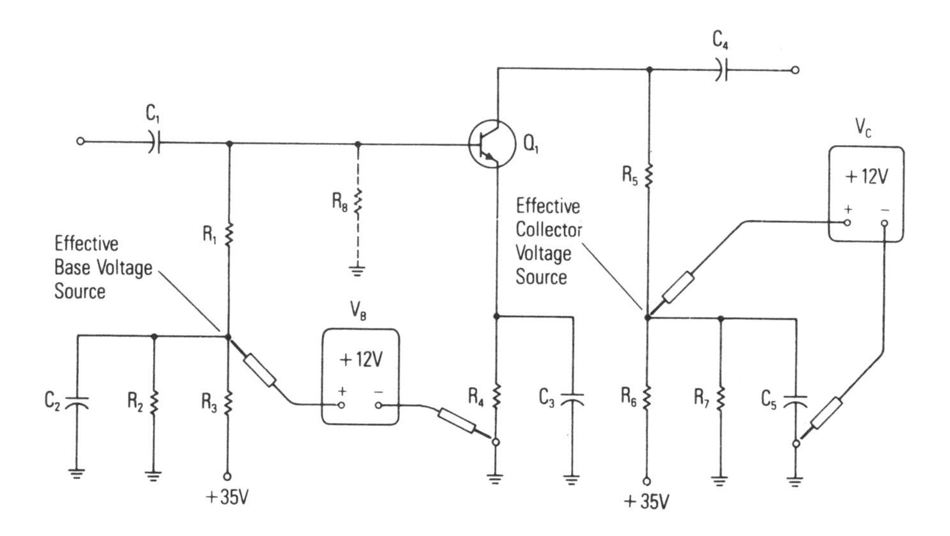

Figure 5-15. Transistor Amplifier Stage with Postive Ground Power Supply

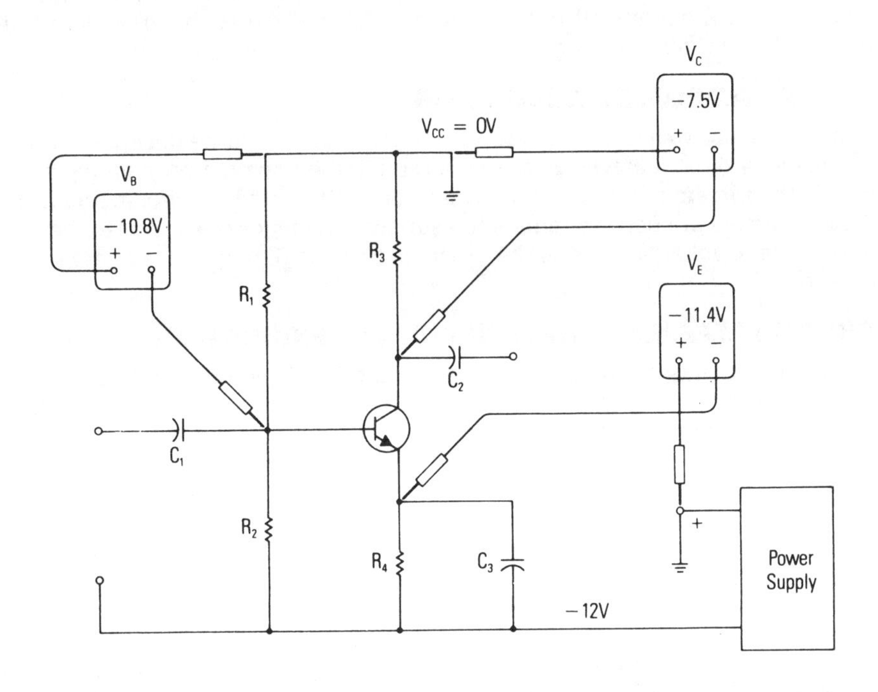

Some technicians prefer to measure the voltage directly between the transistor elements. This is the fastest and least confusing method of establishing whether the bias on a transistor is correct. However, most schematics and service manuals specify voltages as measured from each element to a common bus or ground. Thus, the technician must take the voltage measurement at each element to ground and subtract to find the bias voltage (difference in voltage between the elements). For example, the voltage on the base of the transistor in *Figure 5-15* is −10.8 volts from ground. The voltage on the emitter is −11.4 volts. Thus, the base-emitter voltage is +0.6 volt, the difference between −10.8 volts and −11.4 volts. The base voltage is more positive than the emitter voltage; therefore, the base-emitter junction is forward biased.

SUMMARY

With this chapter, we have completed the individual component measurements and the measurement of components in circuits. In the next chapter, we will look at common types of measurements that normally are made around the home.

CHAPTER 6

Home Appliance Measurements

OBJECTIVE

Appliances in the home used to be very simple. The circuit was made up of a plug for the household wall outlet, the electrical cord, a switch and a load. Most likely, the load was a motor, a lamp or a heating element. Today, however, the circuit for even small, inexpensive appliances may be more than simple. Often the home appliance, such as a coffee maker or vacuum cleaner, will have a microcomputer in it. The microcomputer or microprocessor has assumed an important role in the appliance industry. They are being used as an integral part of many electronic controls for timing, sequencing and controlling events. As a result, testing is likely to be more difficult. Yet, there are many things that can be tested, so that troubleshooting and repair of most appliances is still possible in the components individually and in circuits.

We now wish to illustrate various uses of a multitester by discussing common measurements that you can make on several home appliances to verify proper electrical circuit operation, identify a circuit problem if there is one, and correct or repair any improper operation. We will go through several multitester applications in detail to indicate the procedure. Other applications will not include as much detail, but with schematics and our explanation of the circuit performance, you should be able to use similar procedures to check the circuit performance.

A schematic diagram of an appliance circuit seems complicated and confusing at first glance. It presents many paths for current and many more components than the simple circuit schematics that we have used up to now. The circuit may be a series circuit or a parallel circuit, but in many cases, it will be combinations of both series and parallel circuits. The schematic is a "roadmap" of the circuit and shows the wires, fuses, switches, and current-using devices and their interconnection to form the various paths for current. The schematic actually consists of a number of simple circuits, each with a specific function, combined into the total circuit. To begin to understand such circuits and how to measure them, let's use a common heating and air conditioning system.

HEATING AND AIR CONDITIONING

A heating and/or air conditioning system is one of the most common systems around the home. In many homes, the furnace system to provide the heat is separate from a system to provide cooling. However, more and more homes are being built with a central system that contains both heating and air conditioning. We will use such a system as our example. Heating is provided by a hot-air system which uses natural gas or electricity to provide the heat. Air conditioning is provided by an electrically-powered refrigeration system with the compressor and condenser outside the house and the evaporator inside the house on the output side of the furnace plenum. The furnace fan/blower is used in both cases to distribute the hot or cold air throughout the house.

ELECTRICAL CIRCUIT

Figure 6-1 shows a schematic wiring diagram of a typical central heating and air conditioning system. The thermostat has two switches. SW1 controls the system function — COOL-OFF-HEAT — and SW2 controls the fan/blower — AUTO-ON. In the ON position, SW2 causes the fan/blower to be on continuously regardless of the setting of SW1. In the AUTO position, SW2 allows the system to control when the fan/blower is on.

When the system switch or the fan/blower switch is in the set position indicated by the legend shown in *Figure 6-1,* the contacts are shorted together. FR is a relay that controls the fan/blower motor through contacts FR_1 and FR_2. When these contacts are closed, 115VAC is applied to the fan/blower to cause it to run. CR is a relay that controls the operation of the compressor through contacts CR_1 and CR_2. When these contacts are closed, 220VAC is applied to the compressor motor to cause it to run.

Notice that the control system uses the relatively low and much safer 24VAC that is supplied by the step-down transformer TR_1 from the 110-120VAC line. The 24VAC from TR_1 supplies power to the fan/blower relay, the compressor control relay, and the gas control valve. The latter allows gas to flow into the combustion chamber when heat is required. A gas pilot ignites the gas in the combustion chamber. (In *Figure 6-1,* a heater relay, HR_1, also is shown which controls 220VAC electric heater elements. A real system would have either the gas control or the electric heaters; it is not likely to have both.)

GENERAL PROBLEMS

Normally, when there is an electrical problem in an appliance, only one component is not functioning as it should. For example, common problems might be: the fan/blower fails to come on, or the fan/blower runs but the unit is not heating or it is not cooling. Assuming it is an electrical problem, our job is to find out which of the electrical devices — fan/blower motor, relay, heating element, compressor motor — is not functioning. By looking at the schematic and observing system operation as available controls are changed, you should try to determine which component is not working properly, and the circuit which must be complete to operate the nonfunctioning device. With the help of the schematic, you should be able to quickly determine which wires and components could be causing the problem. Then you can physically locate the component and trace the circuit which must be completed to supply current to the component. Finally, use your multitester as a voltmeter, ammeter or ohmmeter, as required, to determine the cause of the problem.

For example, if the compressor doesn't run or the electric heating element doesn't heat, a simple continuity check will indicate if the compressor or heating element is open. However, before using your multitester as an ohmmeter, all power must be removed from the circuit. *Throw all main power switches on your home distribution panel for the heater and air conditioner to*

Figure 6-1. Schematic Wiring Diagram of Typical Central Heating and Air Conditioning System

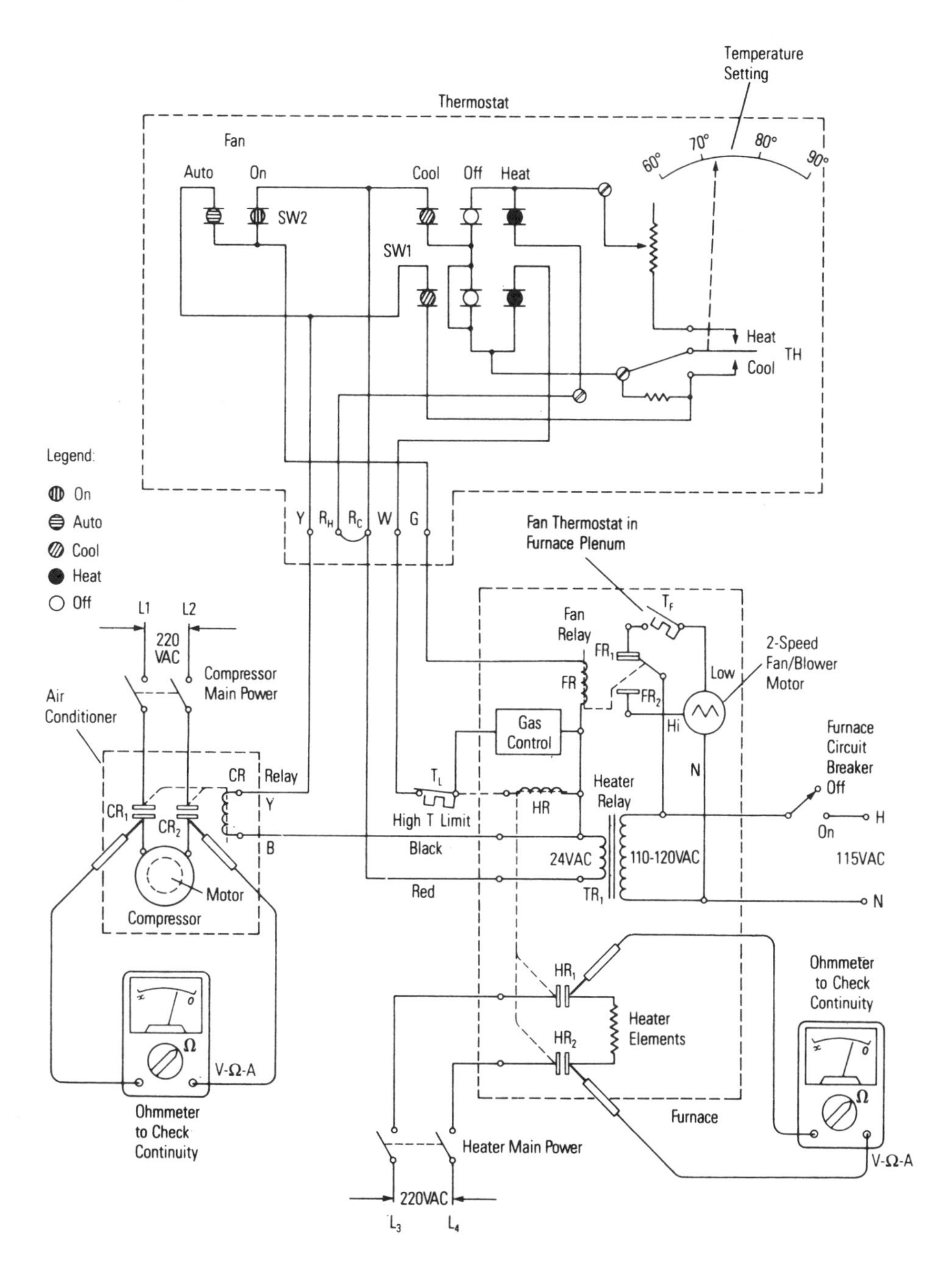

off before making any ohmmeter measurements. Extreme caution should be taken because 220VAC is the power source. Some furnaces also have a power cord plugged into an outlet in the furnace compartment — also unplug this power cord. Before making a continuity check, check with a voltmeter to make sure no voltage is present across the contacts to be measured. Find the CR or HR and make the measurement at the contacts as shown in *Figure 6-1.* No continuity indicates that the windings of the compressor or heater element, respectively, are open.

Isolating the Problem

Most of the problems with the heating or air conditioning system will be in the low-voltage control circuits. Rarely do the relays go out, unless their contacts are burned due to a short in the motor, compressor or the heating element.

The best way to find a problem if the heater or air conditioner does not work is to isolate the system operation into the respective heating and air conditioning functions.

HEAT CYCLE

Figure 6-2 shows the schematic of *Figure 6-1* when in the heating mode. Only the circuitry associated with the heating function is shown. The system function switch is in the HEAT position and the fan/blower switch is in the AUTO position. If the system is operating properly, the thermostat contacts TH_H would be closed because the thermostat was turned up to a temperature higher than the temperature of the space to be heated; thus, demanding heat. Closing TH_H energizes the gas control (or the heater relay) turning on the gas flow which burns and generates heat. When the inside of the furnace plenum reaches the temperature set by the fan/blower thermostat T_F, T_F closes and energizes the fan/blower motor, which delivers heated air to the space to be heated. If the temperature inside the furnace ever exceeds the temperature set by the temperature limit switch, T_L, T_L will open and disconnect the power from the gas control to shut off the gas.

Of course, for a gas furnace, the pilot light must be on to heat the thermocouple, and the thermocouple must be delivering its control voltage (in the millivolt range) before the gas control will turn on and allow gas to flow. So if the furnace is not heating, be sure that the pilot light is lit. If all other control functions seem to be working properly, then the thermocouple may be defective and must be replaced.

If the pilot is on and the furnace does not come on when the thermostat is turned up, follow these steps:

1. Remove the panel from the furnace and the cover from the thermostat.
2. Measure the voltage across the secondary of T_R, the 24VAC low-voltage control transformer, as shown in *Figure 6-2.* If there is no voltage at the secondary, measure the 110-120VAC at the primary. If there is no primary voltage, check the furnace circuit breaker at the power panel. If there is primary voltage and no secondary voltage, turn off power and measure the continuity of the secondary to make sure it is not burned out.
3. If 24VAC is present, manually operate TH so that TH_H closes and determine if gas flows and the furnace flame comes on. If gas does not come on, switch SW_1 from OFF to HEAT several times to determine if the switch contacts

Figure 6-2. Heat Cycle

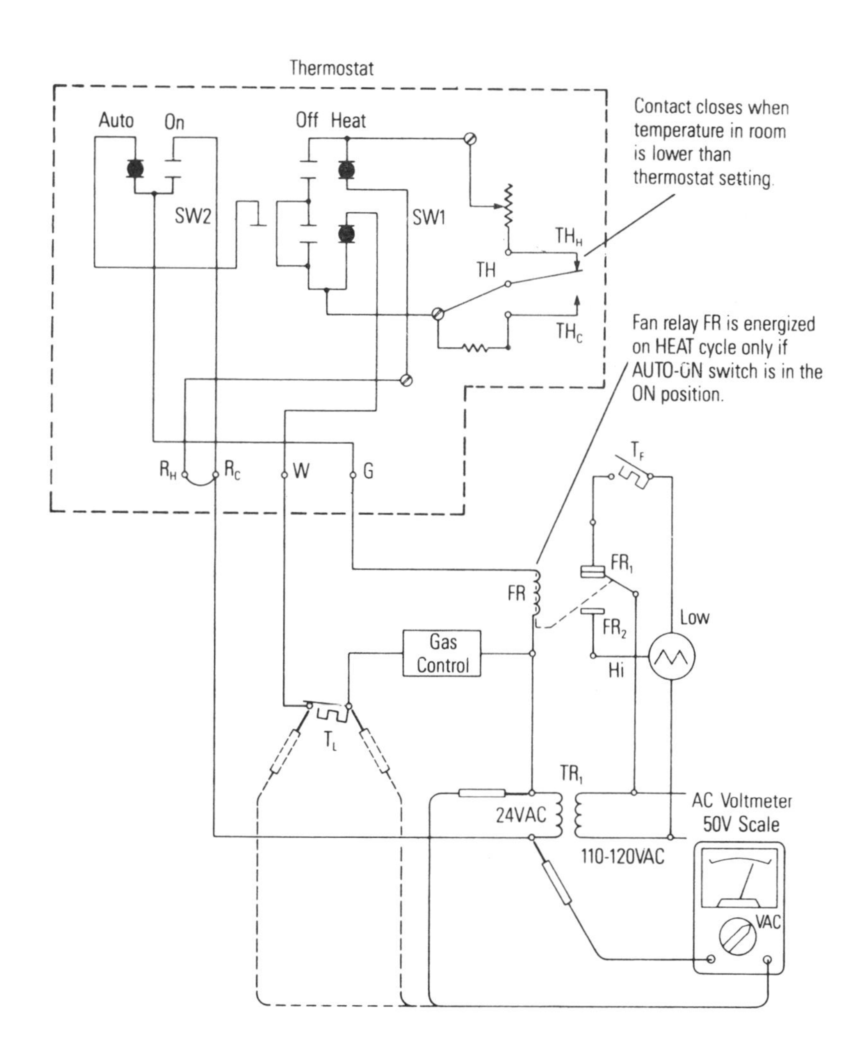

may be corroded and not completing the circuit correctly. Turn off all power and burnish the contacts with a very fine sandpaper. A fingernail file board is a handy tool for doing this.

Another common problem is that the temperature limit thermostat, T_L, is defective. AC voltage measurements around the loop as shown in *Figure 6-2* should isolate the problem.

4. The same procedure can be used to isolate a problem in the fan/blower circuit. However, a first easy check of the fan/blower circuit is to set the fan/blower switch to ON, which will make the fan/blower relay, FR, operate and make the fan/blower run continuously. If the fan/blower does not run,

operate the switch several times. If this does not correct the problem, make ac voltage measurements around the circuit to isolate the problem. 24VAC is much safer to work around, but you should still observe caution; especially be careful that your test leads or tools do not short across the 24VAC source and cause physical damage to the wiring.

The FR relay will make a noise when it operates. If it operates when the fan/blower switch is set to ON but the fan/blower doesn't come on, then there may be problems with the contacts FR_1 and FR_2 of *Figure 6-1*, or the fan/blower motor may be defective.

COOL CYCLE

Figure 6-3 shows the schematic when the system function switch is set to COOL and the fan/blower switch is set to AUTO. The temperature in the space to be cooled is higher than the thermostat setting, thus TH_C is closed to demand cooling. When TH_C closes, it completes the circuit to supply 24VAC to the compressor control relay, CR. The compressor runs and supplies coolant to the evaporator mounted in the air flow path in the furnace.

At the same time the TH_C completes the circuit for CR, it also completes the circuit to supply power to the fan/blower relay, FR, through the fan/blower switch, SW_2, which is set to the AUTO position. This turns on the fan/blower motor and the room air is forced through the evaporator coils to supply cool air to the space to be cooled.

Isolating a Problem

Obviously, if the air conditioning system is not cooling, many problems could be present besides the electrical circuit; therefore, a qualified air conditioning repair person must be called to isolate the problem. However, several simple checks will determine that the electrical circuit is operating properly.

Fan/Blower Motor Circuit

First of all, the fan/blower circuit can be checked easily by placing the fan/blower switch in the ON position. This should make the furnace fan/blower run continuously. As shown in *Figure 6-3*, moving the fan/blower switch to ON completes the circuit to supply 24VAC directly from the secondary of TR_1 to the fan/blower relay, FR. If the fan/blower does not come on, move the fan/blower switch from AUTO to ON several times; the contacts may be corroded and the movement may burnish them to make contact.

If the fan/blower still does not come on, remove the furnace panel and isolate FR and TR_1. First, measure the secondary ac voltage of TR_1 to assure that the transformer is operating properly and supplying 24 volts. Second, measure ac voltages around the circuit as shown in *Figure 6-3* to isolate the problem. The FR relay will make a noise when it operates. If it operates when the fan/blower switch is turned to ON but the fan/blower doesn't come on, then there may be problems with the contacts FR_1 and FR_2 of *Figure 6-1*, or the fan/blower motor may be defective. *With all furnace and air conditioner power off (e.g., circuit breakers off at the power panel,* use your multitester as an ohmmeter for continuity checks to isolate the problem.)

Figure 6-3. Cool Cycle

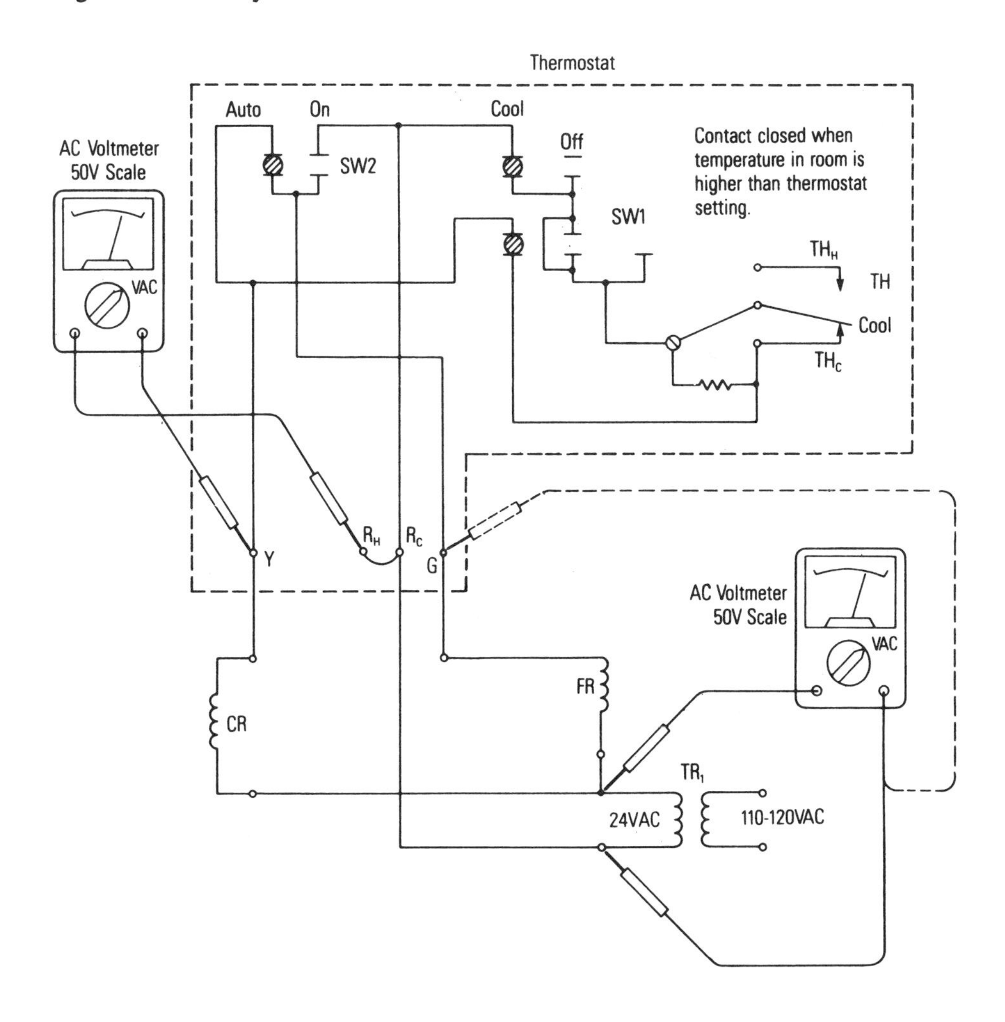

Compressor Circuit

From *Figure 6-3*, we see that the compressor control relay, CR, receives its 24VAC supply through a circuit that is completed through the contacts of SW_1 when the switch is in the COOL position. CR is located in the housing for the compressor and air conditioning condenser that is mounted outside the house or building to be cooled. If the temperature in the room is higher than the thermostat setting, TH_C should be closed and CR should operate when SW_1 is placed in the COOL position. A person located near the condenser can hear CR operate when SW_1 is placed in the COOL position. The compressor should turn on when CR operates. If CR is operating and the compressor turns on, and the fan/blower motor turns on, but the unit still does not cool, an air conditioning specialist should be called.

If CR does not operate, then check to determine that TR_1 is supplying 24VAC properly. An easy check of this is to remove the thermostat cover and measure the voltage at Y and R_C, R_H of *Figure 6-3*. When TH_C is open (no cooling — thermostat set higher than temperature in the room) then the voltage across Y and R_C, R_H should be 24VAC. When TH_C is closed, the voltage should be zero volts. Just with these simple checks you can determine if the electrical circuits are operating properly.

If CR operates but the compressor does not come on, the contacts CR_1 and CR_2 may be bad or the compressor may be defective. Call an air conditioning specialist if the problem is isolated to this point. CAUTION: *If CR is operating properly and the compressor comes on, do not cycle the compressor off and on rapidly. Doing so may burn out the compressor.*

USING A TEMPERATURE PROBE

Some multimeters can be used to measure temperature directly with a temperature probe connected to the meter. The meter shown in *Figure 6-4* has its temperature scale calibrated in degrees Fahrenheit (°F) and in degrees Centigrade (°C). These multitesters actually function as an ohmmeter when measuring temperature. For the analog clamp-on multitester of *Figure 6-4*, set the FUNCTION/RANGE selector switch to the temperature position (X10 TEMP), plug the test leads into the −COM and Ω jacks, short them together, and zero the meter to the CAL position with the ZERO OHMS adjustment. Remove the test leads from the jacks and plug in the temperature probe. Place the temperature probe directly in contact with the surface to be measured for temperature or let it be exposed to the atmosphere being measured. The temperature can be read directly from the temperature scale.

Figure 6-4. Measuring Temperature Directly with Temperature Probe

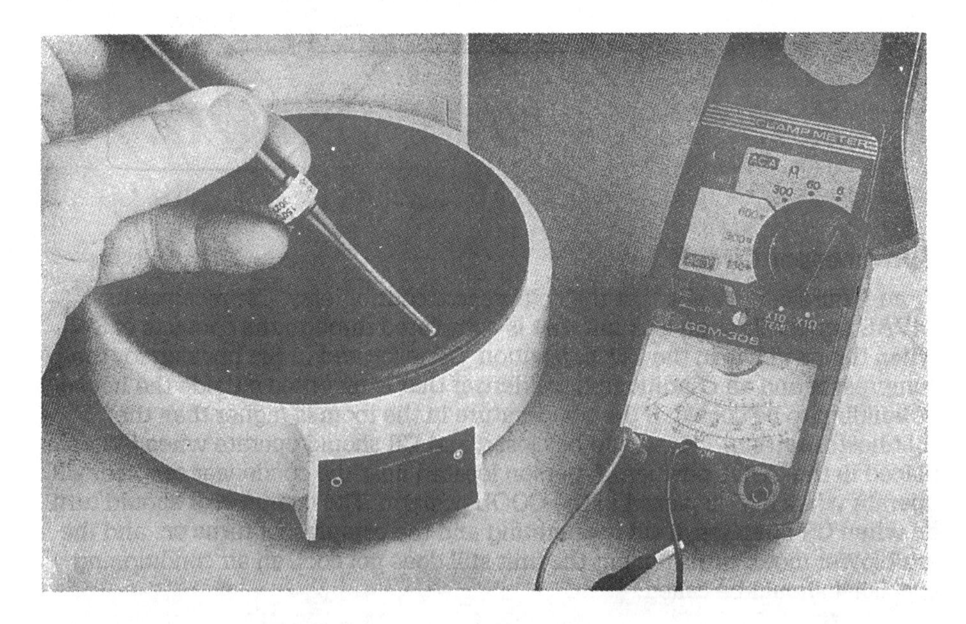

WASHER AND DRYER CIRCUITS

We will not go into detail with respect to locating and isolating problems with the washer and dryer electrical circuits. We will just present the circuit and a short description of the circuit operation and expect that, after the above examples and the material in the preceding chapters, you will be able to follow measurement procedures for continuity, voltage, current and resistance and isolate any problems. To that end the following are some overall guidelines.

Troubleshooting the Electrical System

In almost all cases, troubleshooting the electrical section of a system is straightforward and usually simple. Always apply basic rules of electrical theory when testing any circuit. Remember, if theory and practice do not agree, it means either incorrect theory or incorrect practice!

The first step in troubleshooting any circuit is to have a clear understanding of the circuit and its function before starting. If you do not understand how a circuit functions when there is no problem, it is almost impossible to troubleshoot the circuit because you may not know what you are looking for. This does not mean you must have a total understanding of the circuit, but it does mean that you must have a general overall knowledge of the circuit's function.

The next step is to eliminate the obvious, no matter how simple it may seem. This includes first checking the fuses, circuit breakers, and overload resets. If careful observation does not yield recognition of an out-of-the-ordinary condition, then the problem must be isolated to the control circuit, power circuit, load, or incoming power. Each of these areas, although connected and related, can be isolated from each other to make troubleshooting easier.

The following is a step-by-step procedure:

1. Eliminate the incoming power supply as the source of problems by measuring the 220VAC or 115VAC that is the main supply. One of the most common troubles found in all electrical circuits is a blown fuse or tripped circuit breaker. *Be very careful. Do not touch any metal part of the test leads.*
2. Usually the control circuit of most systems is at a low voltage — 24VAC is common. Isolate it and measure it and make sure it is operating properly. By measuring voltages around a circuit loop, the presence of the 24VAC can be used to indicate continuity through a particular branch of the control circuit. A complete circuit will have current through the circuit components and the measured voltage will decrease as measurements are made from the source around the circuit. An open circuit will have the same voltage reading at each component.
3. Electric drive motors, fan/blower motors, and heaters are common electrical loads. Electric motors are essentially reliable machines and require little maintenance in comparison to the rest of the circuit. Check the voltage at the motor to see if it is the correct level; that is, if it matches the nameplate voltage within 10%. If the voltage is correct at the motor, but it does not run, there may be an internal thermal switch in the motor that has become defective or is open because the motor is hot. One common problem for motors is worn out brushes. Check the brushes and make sure they are OK.

4. The current that the motor draws from the supply is a good indication of its condition and here is a good place to make use of the clamp-on ammeters described in Chapter 3. Remember that the clamp must be around **only one** of the motor leads to measure the current. Connecting around two leads causes the magnetic field of one to cancel the magnetic field of the other. *Figure 6-5* shows the use of a clamp-on ammeter to measure the current of a simple motor.
5. The electrical loads are controlled through relays that are controlled by the low-voltage circuit. Even though low voltage is present and the relay is operating, it does not mean that the contacts themselves are not defective. A closed contact should have almost zero volts across it. A defective "closed" contact will actually be open and have line voltage across it.

WASHING MACHINE ELECTRICAL SYSTEM

The Electrical Circuit

One of the biggest stumbling blocks encountered when servicing an appliance is the variety of wiring diagrams, or even worse, the lack of a diagram at all. Each manufacturer has his own idea of a wiring diagram. Once you have interpreted the wiring diagram and/or the sequence chart, the appliance will be much easier to troubleshoot. *Figure 6-6* shows typical wiring diagrams for a washer. *Figure 6-6a* indicates the typical complete diagram as it would appear on the back of a washer. *Figure 6-6b* indicates the circuitry contained on the timer. We have added notes to *Figure 6-6* to help in the understanding of the diagrams. Many of these would not appear on a typical diagram on the back of a washer. You have to know what is to happen and when it is to happen before you can tell if there is anything wrong. The sequence of events is controlled by a timer.

Figure 6-5. Measuring Motor AC Current with Clamp-on Ammeter

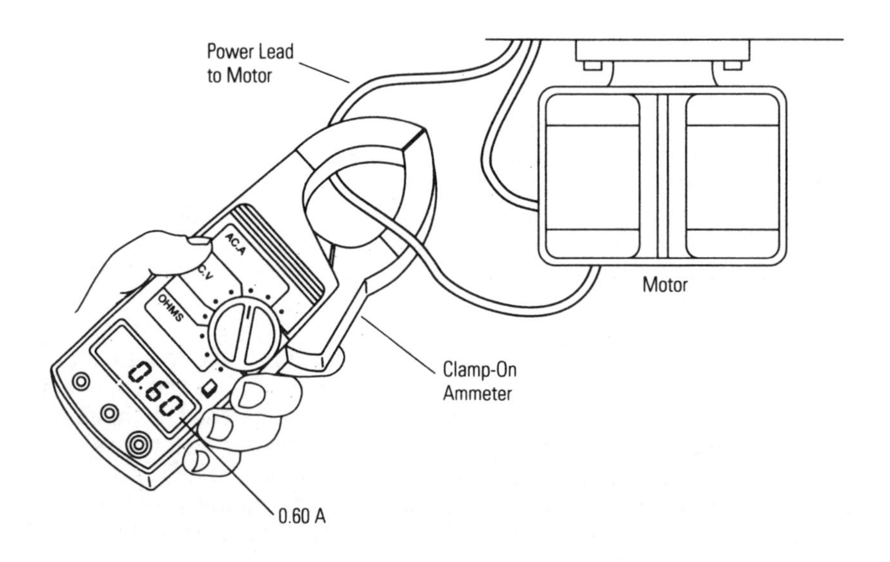

Figure 6-6. Typical Automatic Washer Wiring Diagram
(Courtesy of Sears, Roebuck and Co.)

TYPICAL DRIVE MOTOR

- Has start winding with start capacitor.
- Centrifugal switch disconnects start winding when motor is up to speed.
- Thermal protector opens line if motor overheats.

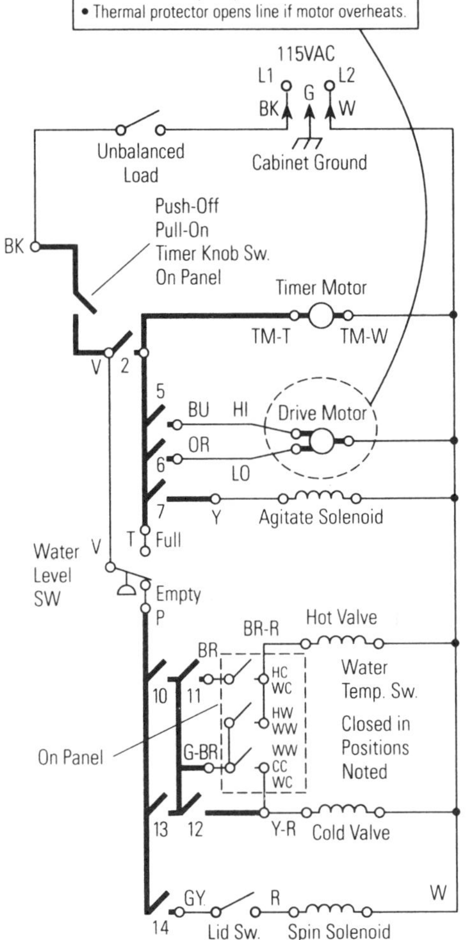

WATER TEMP. SW.

Symbol	Wash	Rinse
HC	Hot	Cold
WC	Warm	Cold
CC	Cold	Cold
HW	Hot	Warm
WW	Warm	Warm

Symbol beside switch means switch is closed for the wash/rinse condition designated.

a. Typical Complete Wiring Diagram

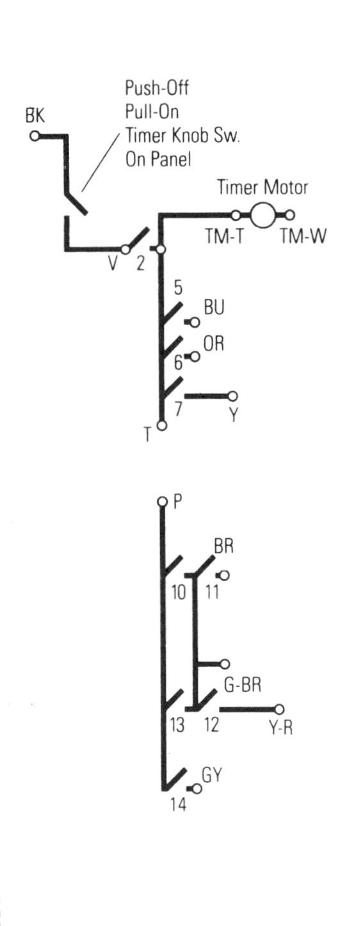

- Small circles indicate terminals.
- Bold lines indicate circuitry is on timer.
- Numbers near switches designate switch's number.
- The number does not appear on timer or other switches.
- GY, G-BK, BR, and etc. indicate insulation color of wire connected to terminal.

b. Typical Timer Wiring Diagram

Timers

The timer in an appliance may be thought of as the "brain" of the machine, because it controls the sequence of operation. As shown in *Figure 6-7,* it generally consists of three basic components assembled into one unit:

1. The timer motor
2. The escapement
3. The cam switchbox.

The motor is similar to that used in an electric clock. It is geared down to a small pinion gear that drives the escapement, which drives cams in the cam switchbox to close numbered switches in a timed sequence. Obviously, if the timer motor is defective, only the motor should be replaced, not the entire timer. The purpose of the escapement is to rotate the cam shaft in a series of timed pulses. It is not generally serviceable and a problem here normally requires replacement. The cams in a timer open and close the electrical switch contacts, thereby controlling the sequence of operations. If contacts are suspected, disconnect power and measure across the contacts with a multitester used as an ohmmeter. Burnish the contacts with fine sandpaper if they are corroded.

A cycle-sequence chart, shown in *Figure 6-8,* tells when a circuit is active, and also at what time in a cycle a particular function is in progress. Be aware that there are delay times between the steps.

Figure 6-7. Typical Appliance Impulse Timer
(Courtesy of Sears, Roebuck and Co.)

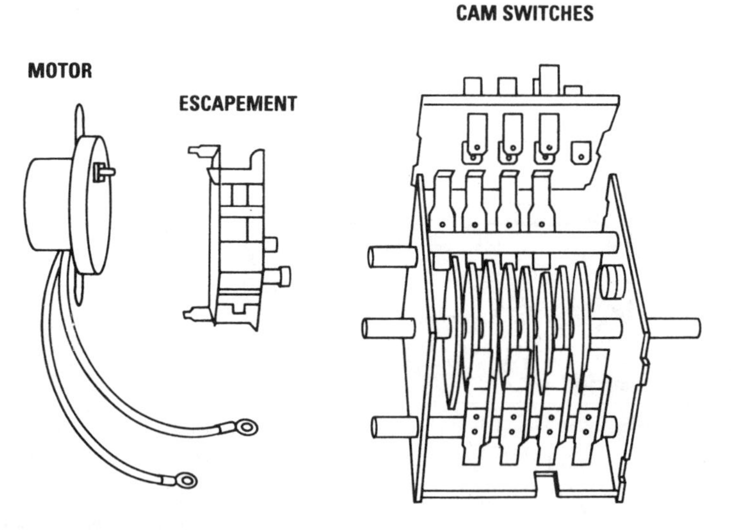

The escapement converts the timer motor's constant motion into stepped movements. The timer moves in set time steps. The time between steps varies from 30 to 120 seconds.

Figure 6-8. Automatic Washer Timer Sequence Chart
(Courtesy of Sears, Roebuck and Co.)

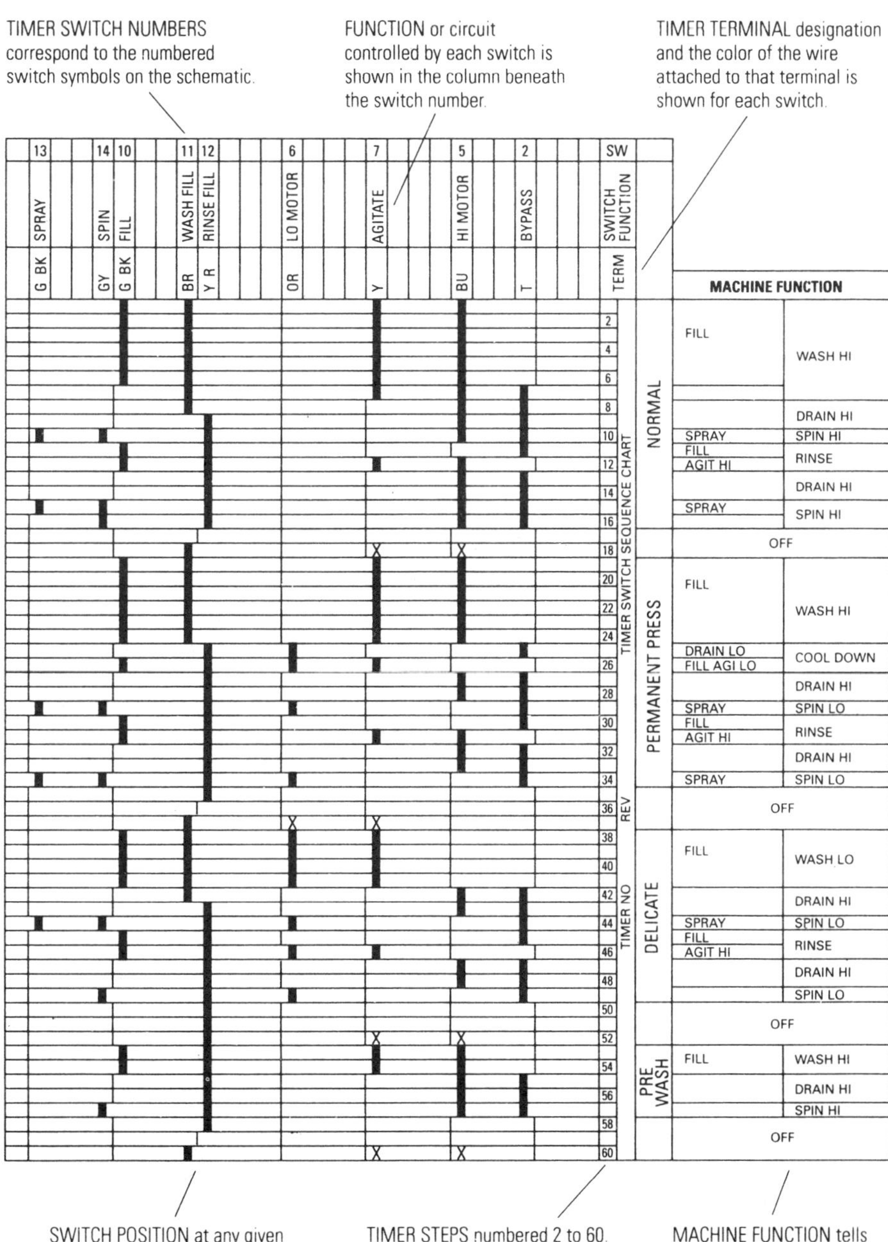

Safety Switches

There are switches that are designed to provide safer operation of the machine, but if they become defective, they may prevent proper operation. An example of such a switch is the one that stops the spin action when the door is opened (a typical one is designated as Lid Sw in *Figure 6-6*). Many machines have a switch that opens the line when an unbalanced load is detected. Thermal protectors are put in motors to protect them if they overheat. Many machines have variable water level switches. These switches are common suspects for open circuits. The water temperature switches on the front panel are not as likely to become defective.

Troubleshooting

With the aid of *Figures 6-6* and *6-8*, you can tell if voltage is to be applied to the timer and drive motors, agitate or spin solenoids, or the hot or cold water valves. If a component is suspected, unplug the machine's power cord and use an ohmmeter to trace the circuit for continuity and to locate accessible terminals where voltages can be measured. Then, if further troubleshooting is needed, plug in the power cord and make appropriate measurements of the 115VAC to verify proper operation. No low voltage control voltage is available in this circuit. Be very careful not to short out the 115VAC.

Here is an example: The washer is on the NORMAL cycle and there appears to be a problem with spinning. Looking at the chart in *Figure 6-8* shows that the normal spin cycle occurs at timer step 10. Moving across the chart at step 10 shows that switches 2, 5, 12, 13 and 14 should be closed. You note that the drive motor is running properly, and that the rinse fill valve is working, but the tub is not spinning. You trace the connections to the terminal on switch 14 of the timer and measure 115VAC with your multitester. You locate the spin solenoid and find zero volts at the spin solenoid. You look for the lid switch and, in the process, find that the lid switch has been broken out of its snap holders and is not being held closed by the lid. You snap it back in place, close the lid, and the spin problem is solved.

DRYERS

The electrical circuit of dryers is very similar to washers in that it contains a timer motor, escapement and a cam switchbox. However, the circuit is much simpler because basically there are only two functions being performed — rotating the tub and blowing hot air through the clothes for a preset amount of time to dry them. The heating element is a resistance wire which glows red hot when a voltage is applied. The voltage is 220-240VAC. The drive motor to turn the tub that tumbles the clothes operates from 110-120VAC. This motor also drives a blower that moves air through the dryer and vents it outside the dryer. The circuit diagram is shown in *Figure 6-9*. This diagram would be in the same form on the back of a dryer as shown in *Figure 6-6* for the washer. We have simplified and rearranged it into a schematic to make it easier to explain. To understand the operation, lets look first at the safety switches.

Figure 6-9. Typical Automatic Dryer Schematic Diagram

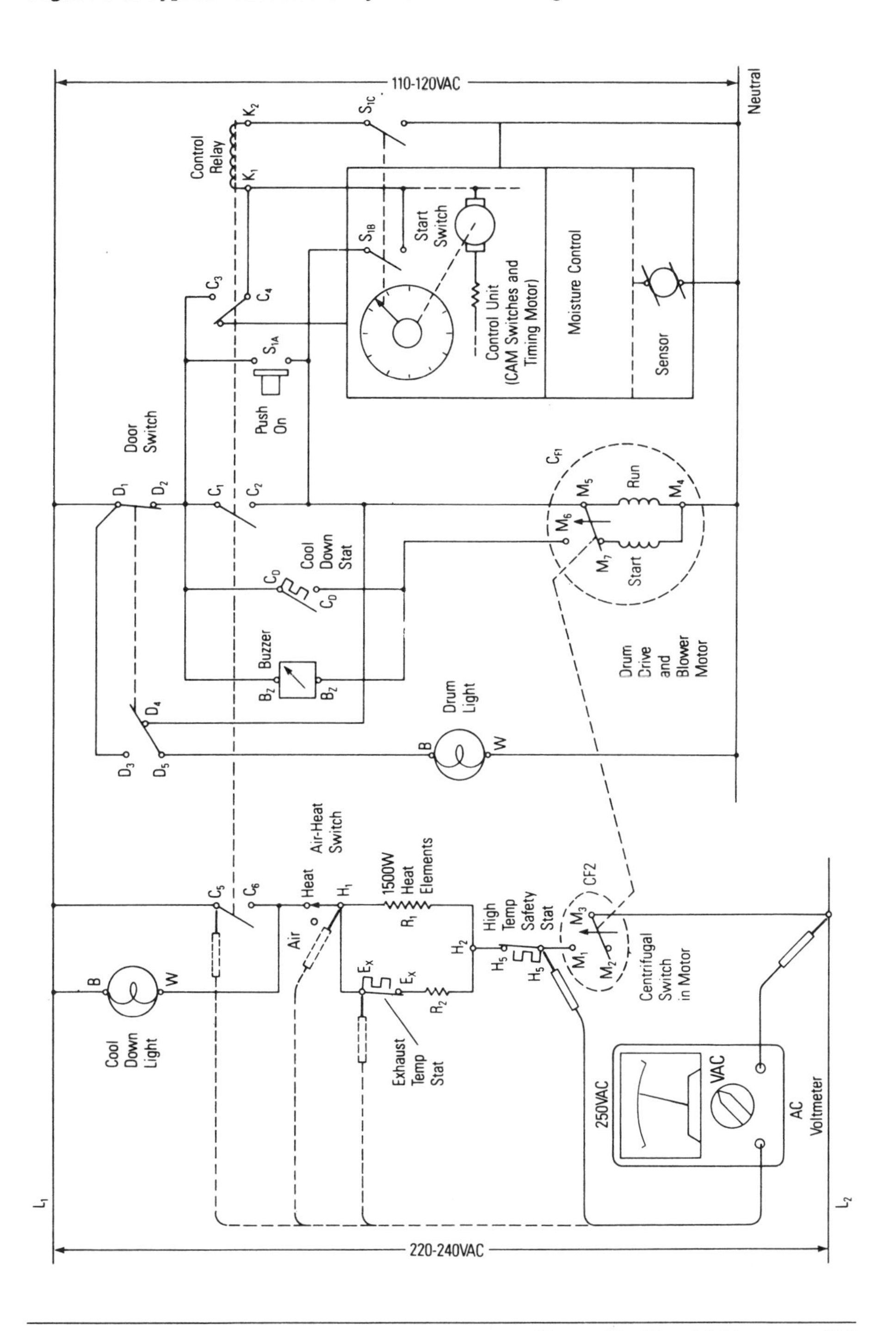

Safety Features

If the drive motor is not turning the tub, there will be no power to the heating elements. The 220-240VAC circuit is held open by the centrifugal switch, C_{F2}, that is on the drum drive and blower motors. C_{F2}, as well as C_{F1}, closes when the motor is at the correct speed. C_{F1} disconnects the start winding of the motor after the motor is at the correct speed. It also closes the circuit to the buzzer and the cool-down thermostat.

The cool-down thermostat closes when the temperature inside the dryer rises to its operating temperature. It does not open again until the temperature cools below this temperature. An exhaust temperature thermostat is located in the exhaust port of the dryer. It is normally closed and will not open unless the exhaust temperature exceeds a given temperature set by the manufacturer of the thermostat. As a double precaution against the dryer rising above a maximum temperature, there is a second safety switch in the heating element power circuit. It is a high-temperature safety thermostat that opens the circuit if the temperature rises above a maximum set by the thermostat. No power will be applied to the drive motor or the control relay circuits unless the door of the dryer is closed. The door switch closes contacts D_1 and D_2 to complete the power circuits. No power could be applied to the heating elements because the control relay would have no power and contacts C_5 and C_6 would be open. All of these switch positions and controls would be presented on a timing sequence diagram similar to *Figure 6-8* for the washer.

Dryer Operation

The start switches, S_{1B} and S_{1C}, are closed by setting the timing control on the control unit to the appropriate time or time cycle desired. The power circuit to the control relay is completed by pressing the momentary push button S_{1A}, which energizes the control relay and closes contacts C_1 and C_2, C_3 and C_4, and C_5 and C_6. C_1 and C_2 complete the power circuit to the drive motor for the drum and blower, and C_3 and C_4 complete a holding circuit for the control relay so it will remain energized after S_{1A} is released. C_5 and C_6 complete the power circuit to the heating elements if the air-heat switch is in the HEAT position. No power will be applied to the heating elements if the air-heat switch is in the AIR position so only room temperature air will be circulated through the clothes.

When the timer times out, S_{1B} and S_{1C} are opened and the control relay is de-energized, which opens contacts C_1 and C_2, C_3 and C_4, and C_5 and C_6. Even though contacts C_1 and C_2 are open, the drive motor does not stop. The cool-down thermostat continues to complete the circuit until the temperature cools below the limit. The power circuit in this case is through the cool-down thermostat contacts C_D and through the M_6 and M_5 contacts of the centrifugal switch in the motor.

Once the cool-down thermostat opens, power is disconnected from the motor. However, until the motor slows down, the M_6 and M_5 contacts remain closed to supply power momentarily to the buzzer that signals that drying is complete. Cool down is indicated by a light bulb that receives its power when C_5 and C_6 contacts of the control relay open. It remains on until the M_1 and M_3 centrifugal switch contacts in the heating element circuit open when the drive motor stops.

The circuitry is straightforward with door safety switches, temperature sensitive switches that turn off the heating element if the temperature of the air over the clothes is too hot, and a centrifugal switch in series with the heating element that keeps the power circuit to the heating element open if the motor is not turning the tub. In other words, the motor that tumbles the clothes must be running before power is applied to the heater.

Troubleshooting a Dryer

A common problem with dryers is that the heating element burns out. With power off (pull the plug from the receptacle in the wall), check the continuity of the heating element with an ohmmeter. This can be done easily by removing the back panel. Check the schematic carefully to determine the proper interconnections.

If safety switches or thermostatic switches appear to be defective, voltage measurements across them should isolate any defective components. Usually if the switch is operating properly, 115VAC will be across its terminals if the switch is open and zero volts if the switch is closed.

Analyze the timing sequence chart carefully when you find the dryer is not working properly. Identify the switches, relays, and motors that should be operating, and troubleshoot the problem in the same manner as demonstrated for the washer.

ELECTRICAL HEATING APPLIANCES

A common appliance around the home is a coffee maker. The coffee maker has two basic functions. First, to heat water to the brewing temperature, deliver the water over ground coffee, and allow the water to drip through the ground coffee into a collecting container. Second, to keep the collected coffee warm after brewing.

A common circuit to accomplish these functions is shown in *Figure 6-10*. The operation is as follows: When S_1 is in the BREW position, 115VAC is applied to the heater through the thermal links, TL_1 and TL_2, and the thermostat, TH_1. The thermostat and thermal links sense the temperature of the water. TH_1 cycles to heat the water to brewing temperature. If all the water is boiled out and the BREW switch is left on, TL_1 and TL_2 become high resistances and restrict the current through the heater to protect the coffee maker from damage.

Once the coffee is brewed, S_1 is switched to the WARM position and 115VAC is applied to the warming coil to keep the brewed coffee warm. At the same time, power is removed from the heater circuit. In the OFF position of S_1, no power is used by the heater or the warmer. Indicator lights show the position of S_1.

Common problems that occur are open heater and warmer elements, burned out WARM and BREW indicators, and open TL_1 and TL_2 thermal links. All of these problems are easy to troubleshoot by using a multitester as an ohmmeter and checking for continuity.

Figure 6-10. Common Coffee Maker Schematic Diagram

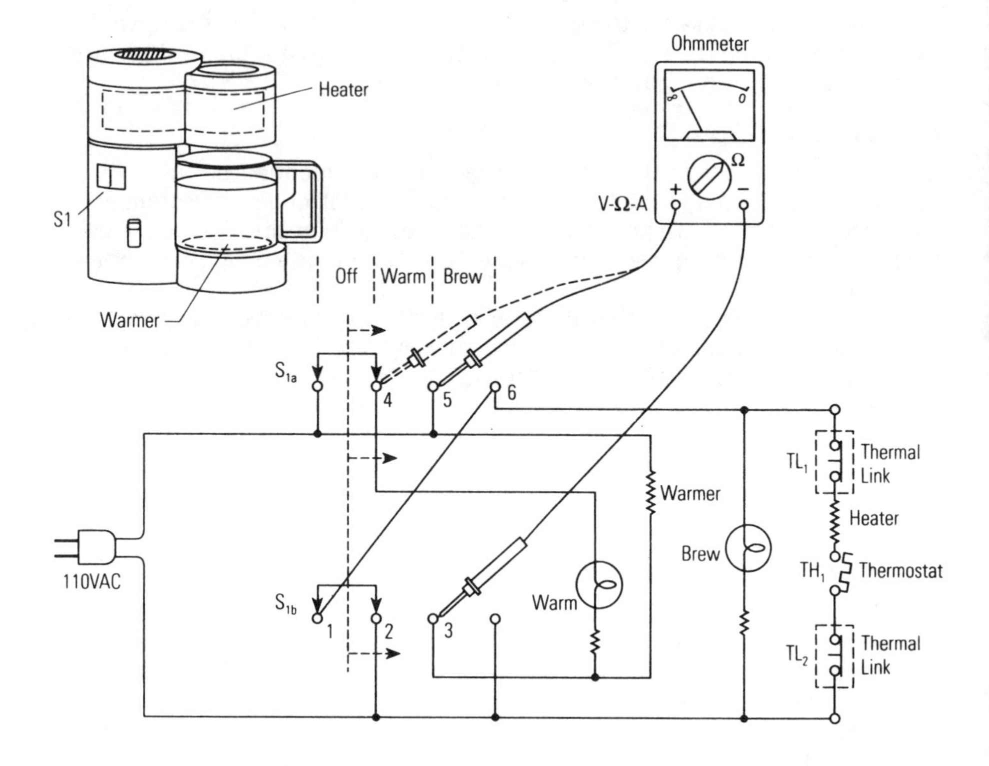

Remove the plug from the 115VAC outlet. Place S_1 in the OFF position. Remove the back cover from the coffee maker to access the connections. Check continuity from pin 4 to pin 3 to establish that the WARM indicator is OK. Check continuity from pin 3 to pin 5 to establish that the warmer element is OK.

Place S_1 in the WARM position. Remove the connections for the heater element and then check across pins 1 and 2 to establish continuity of the BREW indicator. Check across the removed leads of the heater to establish continuity of the heater circuit. There is no need to check voltages in this case, but you can verify them if you want before or after you locate the problem.

SUMMARY

In this chapter, we have looked at finding problems in home appliances. The examples given should provide measurement principles that can be applied to troubleshooting other types of appliances. In the next chapter, we will look at lighting and security systems.

CHAPTER 7

Lighting and Related Systems Measurements

Benjamin Franklin did his kite experiment with electricity in the 1700s and drew many important conclusions which are still used today. However, it wasn't until 1879 when Thomas Edison invented the incandescent lamp (commonly called a light bulb) that the use of electricity became widespread — not only for lighting, but for an abundance of appliances and other devices that plug into the common electric outlet in your home or business. There are many simple measurements and repairs that you can make on your home electrical wiring system. In this chapter, we will look at the more common of this type of measurement.

GENERAL PRECAUTIONS

You don't have to be an electrician, engineer, or have a license to do the installation and testing of simple electrical power circuits. By following the instructions in this chapter, you will be able to do a great deal of them and save expensive repair and installation bills. First, a word of caution concerning two things: (1) All wiring must be in accordance with the National Electrical Code and all local electrical codes; and (2) *the voltage and power in the household wiring circuits can KILL!* Go to your local city electrician to find out about the codes. To reduce the risk of personal injury from electrical shock, fully observe the following simple rules.

Safety Rules

Always observe extreme caution when testing electrical circuits and devices around the home. The high voltages and currents that are present in your home wiring system are dangerous and if you are not careful, they can cause severe damage, injury or death.

- Always be certain that the power has been turned off at the circuit breaker or fuse panel in any situation where you must come in contact with the circuit. Never work on a live circuit unless you are absolutely sure of what you are doing, then proceed with extreme caution. When the power is off, make sure that the power cannot be inadvertently turned on by anyone else.
- Do not stand on damp ground or floors, never in water, and wear dry shoes with rubber soles and heavy socks when working on electrical circuits.
- Even when the power is off, always try to handle wires by the insulation or with insulated pliers — avoid touching a bare wire whenever possible.
- Use only well designed and well maintained tools and measuring instruments to test and repair electrical systems and equipment. Always hold the test probes by the insulated part. Never touch the metal probe tip when testing outlets or circuits.

HOME ELECTRICAL CIRCUITS

Unless your home was built over half a century ago, the electrical power feed to your home has three wires which come from the power company's line transformer. One of the wires is the common or neutral. The other two wires are "hot" and each of them has 110-120VAC with respect to the common line. The voltage across the two "hot" wires is 220-240VAC. The three wires come through the power company's meter to the main power panel in your home.

Main Power Panel

The main power panel is usually a large metal box containing a main power switch and circuit breakers (or fuses) as indicated in *Figure 7-1*. In the main power panel, the main line is divided into a few 230VAC branch circuits and many 115VAC branch circuits. Only one 230VAC branch circuit and two 115VAC branch circuits are shown in *Figure 7-1* as examples. Each branch circuit is protected from overload by either a fuse or a circuit breaker which opens the circuit when there is too much current. Although a fuse or circuit breaker does the same thing, a circuit breaker is better because it can be reset after being tripped. A "blown" fuse cannot be used again, but must be replaced. The main line's common wire is connected to the ground bus in the main power panel. The white wire for each branch circuit is connected to this ground bus. The black wire for each branch circuit is connected to the protected output side of a fuse or circuit breaker.

Division of the electrical load in the home has two main purposes: (1) It keeps the current in each branch low (less than 15A) so the branch wires are smaller, less expensive, and easier to install in the home; and (2) an overload on one branch is isolated so it does not affect other branches.

Lighting Circuit

Circuit #1 in *Figure 7-1* is a lighting circuit which has two 100W light bulbs wired across (in parallel with) Leg 1 of the 110-120VAC main power. A 15A circuit breaker is in series with the black wire of this branch. The wires (pigtail leads) from the light bulb sockets are connected to the lighting circuit by stripping insulation from the ends of the wires and twisting the stripped ends together. Quick connectors, called wire connectors, screw onto the twisted wire ends. The wire connectors hold the twisted wires in place and insulate the bare wire ends. These wires are tucked under the light socket and into the junction box in the ceiling or wall, then the light socket is fastened to the junction box.

To measure the voltage at either of the light sockets, open the circuit breaker for the branch, remove the socket from the junction box, pull out the wires, and remove the wire connectors. Close the circuit breaker and measure the voltage at the twisted wire connections with an ac voltmeter on the 150VAC range or greater. Additionally or alternatively, the voltage can also be measured at the light switch terminal as shown, if it is accessible.

WARNING

Be sure that the main power switch is turned off before making the following connection. Be sure there is no voltage at Point A (Figure 7-1) before inserting an ammeter.

Figure 7-1. Home Power Distribution System

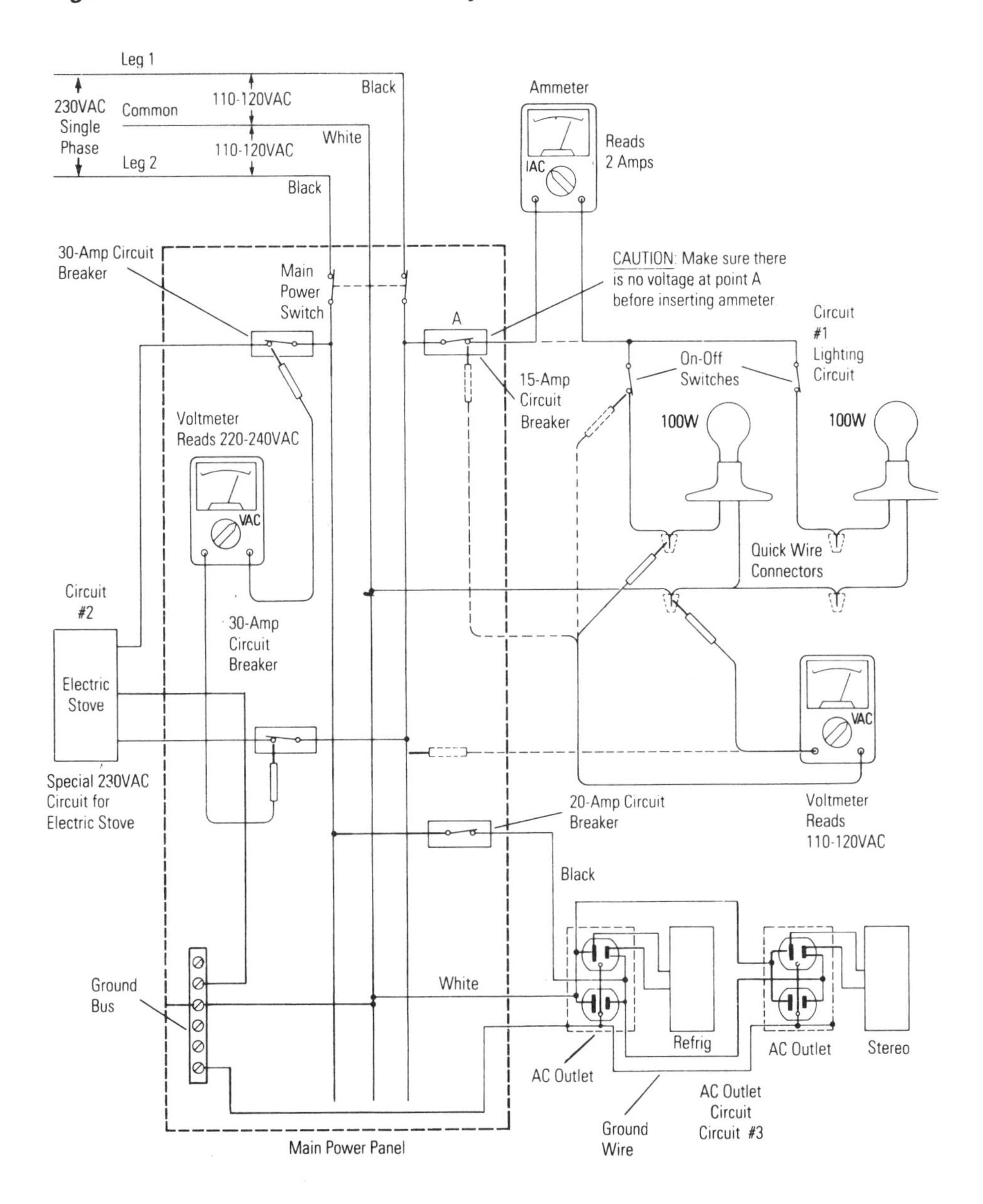

If the current is to be measured and a clamp-on ammeter is not available, the circuit can be opened at the circuit breaker by removing the black lead from the circuit breaker and inserting an ac ammeter in series as shown in *Figure 7-1*. With two 100W bulbs in the circuit, the ammeter should read approximately 2A.

230VAC Circuit

Circuit #2 is a 230VAC circuit for an electric stove. It is connected across both incoming power legs. As shown in *Figure 7-1,* the voltage can be measured easily by placing an ac voltmeter (scale 250VAC) across the two circuit breakers (one in each leg) in the power panel. As a result, the measurement is much safer because no wires or wire connectors need to be removed.

As mentioned earlier, the center wire (between leg 1 and leg 2) is called "common." It is also sometimes called "ground." Technically, there is a difference between the two terms; however, there should be no voltage difference between the "common" and "ground" wires. We will have more to say about "ground" as we discuss the outlet circuit.

115VAC Outlet Circuit

Circuit #3, an ac outlet circuit, is shown in *Figure 7-1* with a refrigerator and a stereo plugged into the outlets. If it is suspected than an ac outlet is not functioning properly, the ac voltage at the receptacle can be measured as described in Chapters 1 and 2. If it has been determined that an outlet does not have voltage, further measurements can be made as shown in *Figure 7-2* by following these steps:

1. If possible, identify another outlet receptacle on the same circuit. Plug in a lamp. Turn on the lamp.
2. Remove the center screw and face plate of the suspected outlet receptacle.
3. At the main power panel, throw the circuit breaker for the outlet circuit to OFF. This should turn off the lamp to indicate that power is off to the circuit.
4. Remove the screws that hold the suspected outlet receptacle in the box.
5. Pull out the receptacle as shown in *Figure 7-2.*
6. At the main power panel, throw the circuit breaker to ON.
7. Measure the voltage at the receptacle's screw terminals as shown in *Figure 7-2.*
8. If there is no voltage, look for broken wires, loose screws, a damaged outlet, and loose connections on "daisy-chained" outlets on the same circuit. If the outlet receives its power by wires inserted into spring contacts in the receptacle rather than the wire being wrapped around a screw, carefully check that the wire is firmly gripped by the spring contacts.
9. If the outlet is OK, continue to isolate the problem by tracing the circuit back towards the power panel. Loose connections at the previous "daisy-chained" outlet or in a junction box in the circuit may be the cause of the trouble. Proceed carefully by turning power ON and OFF as necessary.

Importance of Grounding

Outlets in many older houses and buildings are not properly grounded. Here is why a grounded outlet is important: Suppose an appliance develops an internal short from the 115VAC or 230VAC line to its ungrounded metal case. This causes a very dangerous condition. If you touch the appliance case with one hand, and with the other hand you touch something, such as a plumbing fixture, which is grounded, there is a complete circuit through you. You could receive a fatal shock. However, if the appliance case is grounded via a properly grounded outlet, the excessive current from the shorted "hot" wire to the

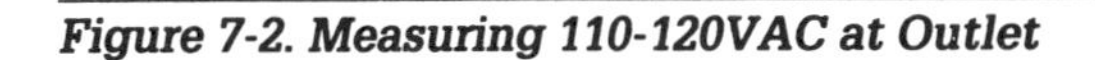

Figure 7-2. Measuring 110-120VAC at Outlet

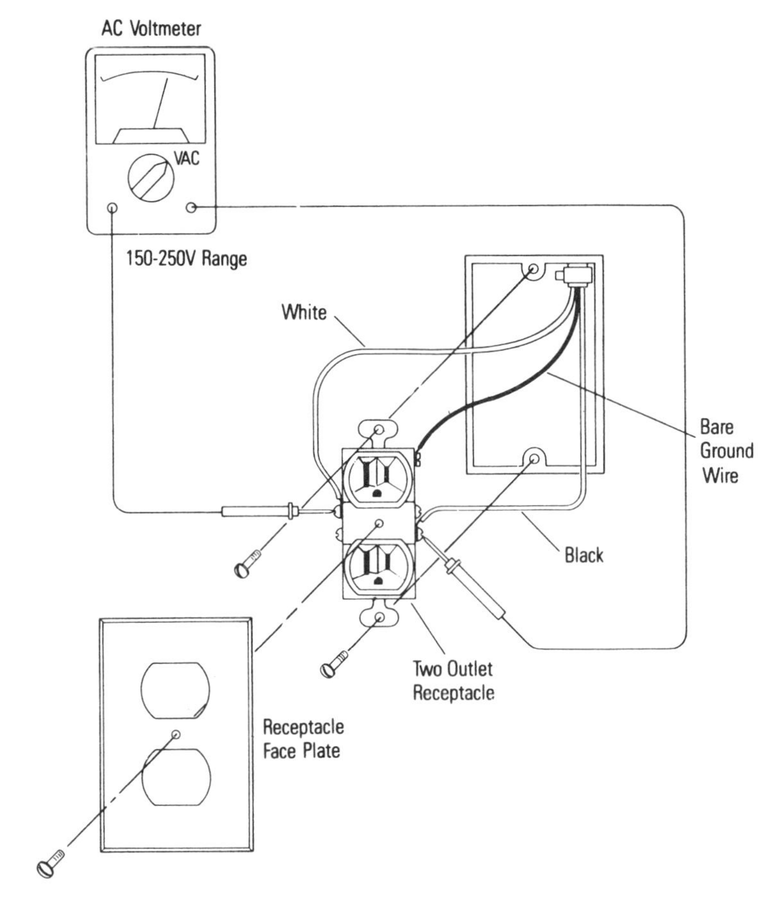

common through the grounded outlet will trip the circuit breaker. Thus, power will be removed from the outlet until the defective appliance is unplugged and the circuit breaker reset. Refer to *Figure 2-4* to find out how to determine if your outlets are grounded.

If you have two-wire ungrounded outlets, you may want to have them replaced with three-wire grounded outlets by a licensed electrician. If you have two-wire grounded outlets, then you may want to replace them yourself, or temporarily install a ground adapter plug as shown in *Figure 7-3*.

Shorts and Ground Faults

A short circuit occurs when current takes an accidental path short of its intended circuit. It is caused by the creation of a path of significantly lower resistance than that of the normal circuit. A short circuit can be damaging to an electrical system because of excessive currents that occur under a short circuit condition. Short circuits are usually caused by insulation between parts breaking down or wires shorting together because of vibration.

Figure 7-3. Installing a Ground Adapter Plug

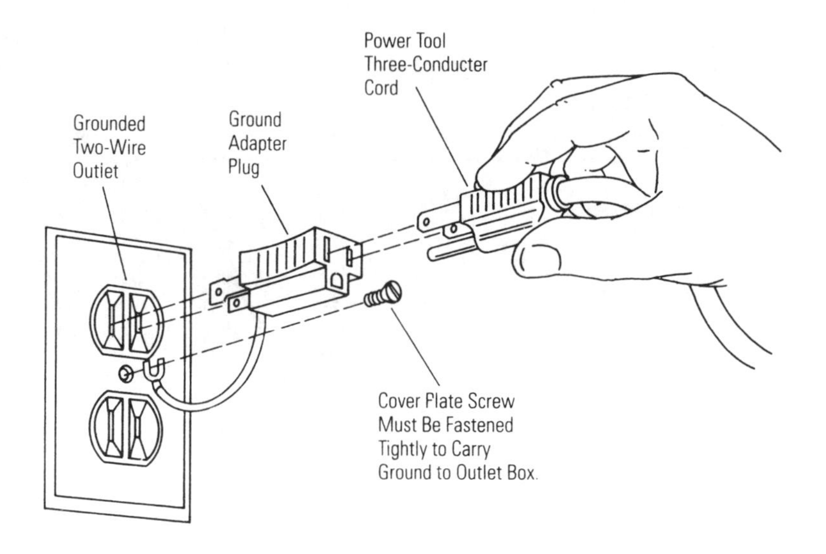

A ground fault is an accidental connection of a non-grounded conductor to the equipment frame or case. It is a form of a short circuit that results in current leaving the circuit conductors and going through a path of other conducting materials provided by the frame. It is usually a fault in insulation.

Both short circuits and ground faults can be located with an ohmmeter. Check the circuit schematics of the equipment under test to determine the normal resistance in the circuit.

CAUTION

Be sure that the power is off before measuring ohms!

Incandescent Light Bulb

The incandescent light bulb is basically a simple device whereby the electrical current passing through the filament wire heats it so hot that it glows and gives off light. The amount of heat produced by the filament is determined by the wattage of the bulb. The wattage, in turn, is determined by the voltage, current, and resistance according to Ohm's law. The quickest and easiest measurement to make on a light bulb just to determine if it has continuity is to measure the filament resistance with an ohmmeter. However, as we pointed out in Chapter 2, the filament's cold resistance is greater than its hot resistance, so that a simple measurement with an ohmmeter will not be correct for finding the wattage. As discussed in Chapter 2, the hot resistance can be found by measuring the voltage across the bulb and the current through it, and dividing the voltage by the current to obtain the resistance.

Dual-Filament Incandescent Light Bulbs

A popular type of table or floor lamp uses a three-way incandescent light bulb which permits the lamp to be set to any one of three brightness levels: dim, medium or bright. The light bulb has two filaments and the lamp fixture has a special switch that connects the filaments singly or in parallel. The continuity of a three-way incandescent light bulb can be checked with an ohmmeter just like the bulb in *Figure 2-7*, except there will be a separate ring in the base for the second filament. If the lamp fixture is suspected to be defective, you can measure the voltage from line to switch with an ac voltmeter.

Fluorescent Lamp

The fluorescent lamp, which was introduced in 1939, is much more efficient than the incandescent light bulb. As shown in *Figure 7-4*, fluorescent lamps or fixtures require a starter circuit to heat the filaments until they emit sufficient electrons to ionize the mercury vapor gas inside the fluorescent lamp. After the gas is ionized, the starter circuit removes heating power from the filaments.

Figure 7-4 Fluorescent Lamp Circuit

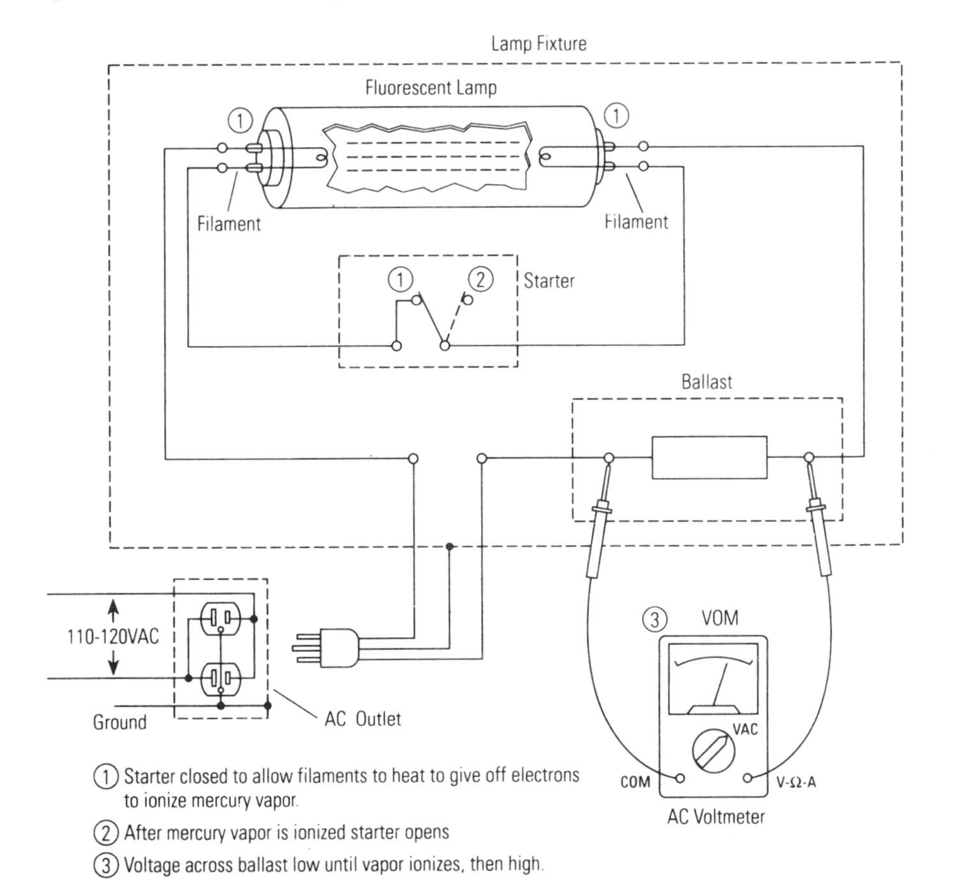

The internal resistance of the lamp drops very low when the gas is ionized and current flows through the tube. If fact, the resistance is so low that if it were connected directly to the power line, it would draw so much current that it would burn out very quickly. A current-limiting device, called a ballast, is connected between the lamp and the power source to limit the current.

In the fluorescent lamp circuit shown in *Figure 7-4*, the starter must close to start the lamp, but on the other hand, if the starter does not open, it will not allow the fluorescent lamp to light. The ballast and starter usually are contained within the fluorescent lamp's fixture. Some early fixtures had removable starters, but now the starter is wired into the fixture. Measuring the voltage drop across the ballast with a voltmeter as shown in *Figure 7-4* gives an indication of the circuit operation. The lamp, the starter, and the ballast are the most likely components to fail, in that order.

LOW-VOLTAGE LIGHTING

A decorative low-voltage outdoor lighting system can provide a safe and economical way to illuminate dark steps and sidewalks. In addition, the low-voltage outdoor lighting is also far more economical to operate than ordinary household lighting fixtures. In fact, the six-light system shown in *Figure 7-5* uses less electrical energy than a single 75-watt light bulb.

The control box contains a programmable timer, a transformer, a circuit breaker, and a test switch to bypass the timer to test the lights. The transformer is what makes this system electrically safe. It converts the primary 115VAC line voltage to a secondary voltage of between 8VAC and 24VAC. The transformer's low-voltage output is fed to two screw terminals on the control box. Incoming power is supplied either through a standard three-prong plug from an ordinary 115VAC outlet, or, as shown in *Figure 7-5*, it must be connected with wire connectors to a 115VAC power line in a junction box. The ground wire (usually green or bare) connects the control box housing to ground.

WARNING

The 115VAC in the primary circuit is very dangerous and the wires must be handled with extreme care as we discussed earlier in this chapter under the heading "Safety Rules."

A heavy gauge (usually #12 to #16 stranded) waterproof two-conductor cable is used for the lights. Some lights have screw terminals while others have contacts to pierce the cable as shown in detail in *Figure 7-5*. The cable can be buried in the earth as it runs between lights. The secondary has no ground connection and, since it is ac, there is no need to be concerned with polarity. The control unit is a clock-timer that operates a switch at the programmed times to turn the lights on and off.

The usual symptoms of a problem in a low-voltage lighting system are either: (1) dim lights (2) all lights out, or (3) one or two lights out.

(1) Dim lights mean extra voltage drop caused by an overloaded line or by bad connections. Using an ac voltmeter as shown in *Figure 7-5*, measure the voltage at the control box and at the end light station. If the voltage difference is greater than two volts, the connecting cable wires are too small. A larger cable must be used. If the voltage drop is small, examine all connections and make sure they are clean and piercing the cable properly.

Figure 7-5. Low-Voltage Lighting System

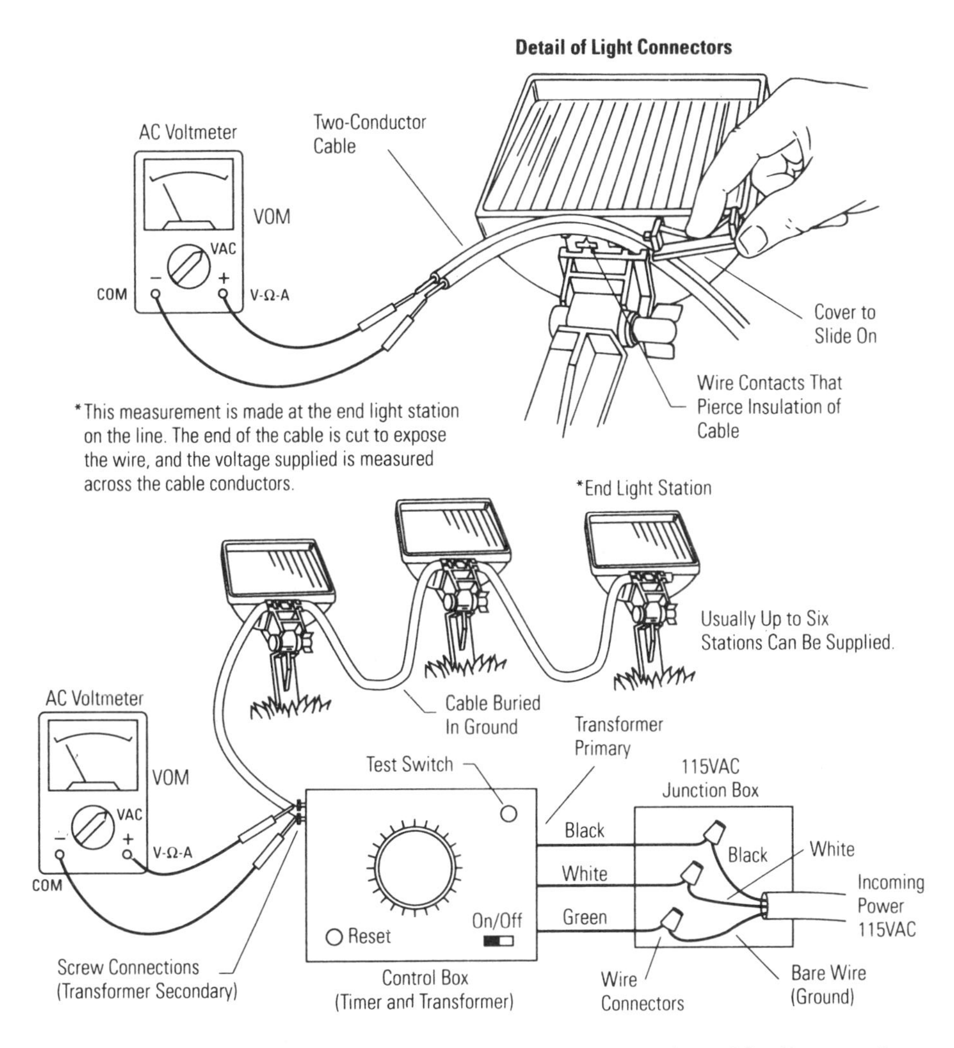

(2) All the lights out means no power is getting to the cable. Be sure the ON/OFF switch is on and the circuit breaker is on. Use an ac voltmeter and measure the voltage at the output of the control box. It should be about 10 volts for a 12V system. Check the 115VAC source if there is no voltage. If there is 115VAC supplied but no output voltage, press the test switch and measure the output voltage again. If there is output voltage, the control box clock-timer is not working properly. Contact your qualified maintenance center for help.

(3) If only one or two lights are out and the others are operating normally, the control box is OK, but there probably are burned out bulbs, bad connections, or faulty wiring. Use your multitester as an ohmmeter or voltmeter to check the bulbs and the cable voltage and this should locate the problem. A good bulb will have a cold resistance of about one ohm.

OPTICAL SENSORS

Optoelectronics is the technology that combines optics and electronics to change light into electricity. Of the many optoelectronic devices, the two simplest are most likely to be found around the home. They are the photoconductive and photovoltaic cells.

Photoconductive cells or light sensors are basically light-sensitive resistors. That is, their conductivity depends on the incident light on the cell as shown in *Figure 7-6.* Many materials are photoconductive to some degree; however, the commercially important ones are cadmium sulfide, germanium, and silicon. The spectral response of the cadmium sulfide cell closely matches the human eye, so it is often used in applications where human vision is a factor. Controlling lights, like the low-voltage system, and automatic iris controls for cameras are two typical applications.

Figure 7-6c shows how the photoconductive cell is used to turn on a light at dark and off at dawn. To test the photoconductive cell, it must be removed from the circuit and measured with an ohmmeter. When you cover it from light, its resistance will be very high — greater than 100,000Ω. When illuminated, the cell's resistance may be less than a few hundred ohms.

Whereas photoconductive cells can only control current, photovoltaic cells are devices which produce a voltage when they are exposed to light. A popular photovoltaic cell is a solar cell which converts the radiant energy of the sun into an electrical power source. In many applications, such as calculators, there is no other external power source but the photovoltaic cells. The photovoltaic cell's output may be measured with a dc voltmeter as shown in *Figure 7-7.* When illuminated with bright sunlight, a single cell's output voltage may be as much as 0.55 V. They can be connected in series for higher voltage and in parallel for higher current capability.

Figure 7-6. Photoconductive Cell

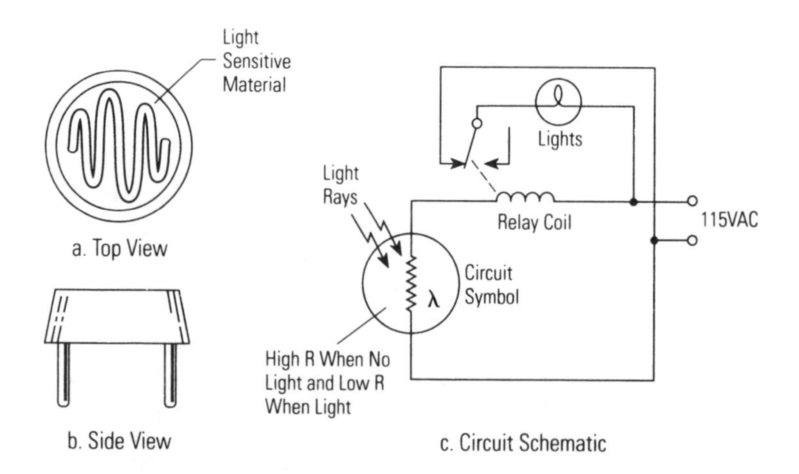

Figure 7-7. Photovoltaic Cell

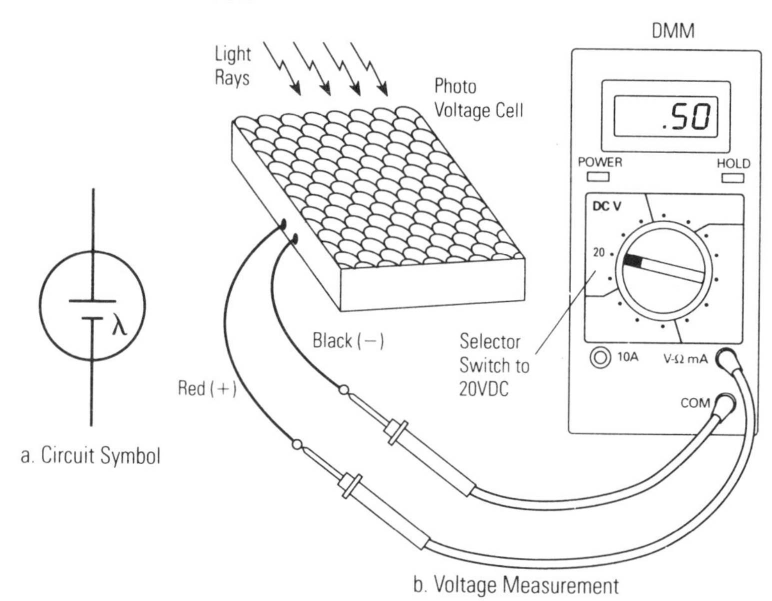

GARAGE DOOR OPENER

The radio-linked electronic control of a garage door opener is very sophisticated with security systems involving digital codes (or channel selection) and output signals to activate lights both inside and outside of the house. The circuitry outside of the transmitter and receiver, however, is simple enough to allow you to perform troubleshooting with your multitester. The basic circuit is shown in *Figure 7-8* with the most likely test points indicated. The relay that operates the motor can be activated either by the radio transmitter or the hard-wired push-button switch mounted on the inside of the garage.

Locating a Problem

If the push-button switch activates the motor control relay, but the transmitter does not, the first thing to check is the battery in the transmitter by measuring it as described in Chapter 1. If the battery is good or if you replace it, and the remote transmitter still does not activate the motor relay, check the digital code or channel selector switch to make sure receiver and transmitter are on the same code or channel. Check the owner's manual to determine if there is a test point at the receiver (as shown in *Figure 7-8*) to measure with a dc voltmeter an output voltage that indicates if the transmitted signal is being received. In many cases, the receiver, when it receives a transmitted signal, shorts out the same terminals (1,2) as the push-button switch; however, in *Figure 7-8* the receiver has its own inputs (3,4.) If the indication is that the transmitter or receiver is at fault, refer their servicing to a qualified service technician.

If the push-button switch does not activate the motor relay, you can bypass the push-button switch circuit by shorting across the screw terminals 1 and 2 with a screwdriver, as indicated in *Figure 7-8*. If this shorting activates the motor, unplug the control from the 115VAC outlet, use your multitester as

Figure 7-8. Garage Door Opener

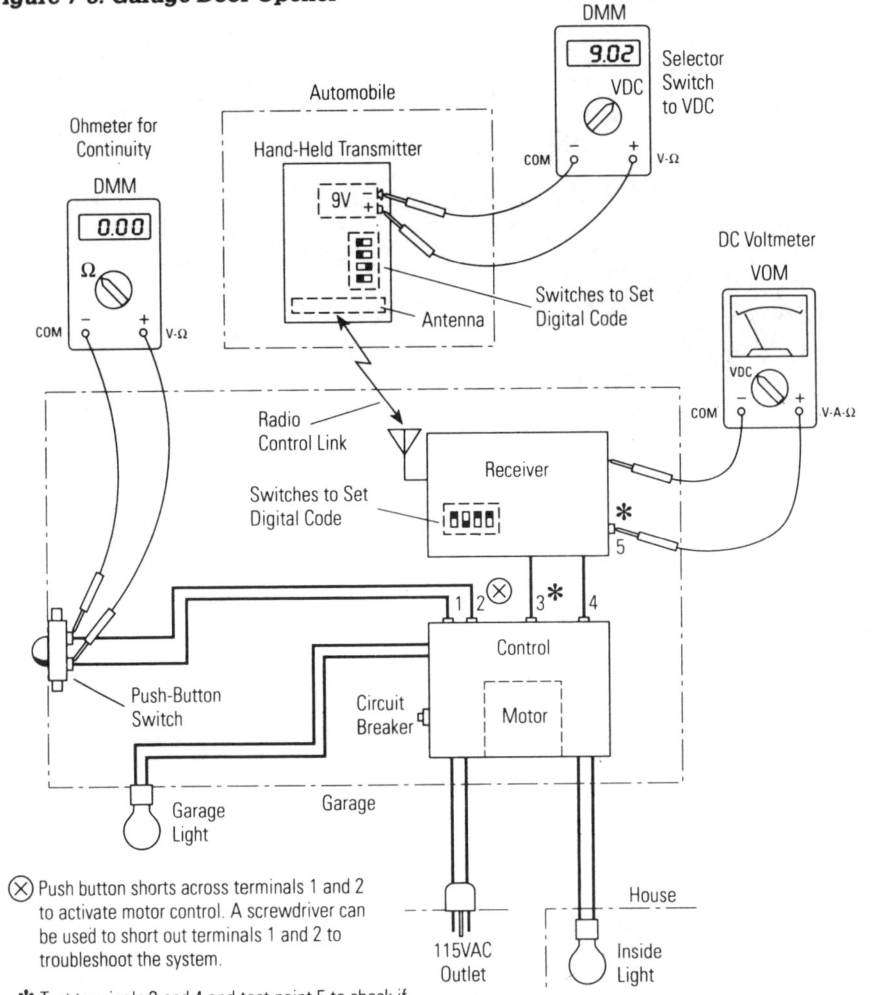

⊗ Push button shorts across terminals 1 and 2 to activate motor control. A screwdriver can be used to short out terminals 1 and 2 to troubleshoot the system.

* Test terminals 3 and 4 and test point 5 to check if receiver is being activated by transmitter. Receiver output essentially does the same thing as push button to activate motor control.

an ohmmeter and check for broken wires to the push-button switch and continuity problems in the push-button switch itself.

If the motor is not activated, measure with an ac voltmeter to be sure there is 115VAC power to the unit. Many units have a circuit breaker in them. Make sure it has not been activated. If it has, reset it. The shorting of terminals 1 and 2 usually causes a stepping relay in the control box to activate. If the stepping relay activates but the motor doesn't, check the contacts of the stepping relay. They may be burned or shorted together. If they are shorted, the circuit breaker has probably been activated. If none of these checks reveal the trouble, you may have to call a service technician.

INSTALLING A CEILING FAN

There are two ways that a ceiling fan can be physically installed: (1) The traditional installation using the down pipe *(Figure 7-9a)*, or (2) the close to the ceiling, low profile installation *(Figure 7-9b)*. Regardless of the physical type of installation, the electrical wiring is the same except for running the wires through the down pipe. Typical fan kits are shown in *Figure 7-9*.

To reduce the risk of electrical shock, turn off the power at the circuit breaker or fuse box before beginning. An assortment of light kits are usually available to attach to the ceiling fan. The connections with wire connectors of the blue-to-black and white-to-white wires between the fan and light are shown in *Figure 7-9*. The fan and the light each have a pull chain to operate their individual internal switch to control the fan speed and light, respectively. If this provides all of the control that you want, wire the fan as shown in *Figure 7-9a*. The black wire from the fan, the blue wire from the light, and the black wire from the 115VAC branch circuit are all twisted together and fastened with a wire connector. The common white wire from the fan and light and the white wire from the 115VAC branch circuit are twisted together and fastened with a wire connector. The bare ground wire from the branch circuit should be connected to the outlet box. Connect the green ground lead from the fan to the outlet box as well.

Some people want to have additional control, such as a wall switch for the fan and/or light, or a solid-state dimmer for the lights. The circuit shown in *Figure 7-9b* adds a wall switch or dimmer for the light, and a fan switch. The white wire and ground connections are the same as for *Figure 7-9a*. The black wire from the fan and the blue wire from the fan light are connected through their respective switches (or dimmer) to the black 115VAC supply wire. Do not connect a solid-state light dimmer to the fan motor. It will cause the fan to overheat. If a fan switch is not required, then the black wire for the fan is connected directly to the black 115VAC supply wire as indicated in *Figure 7-9b*.

The wires that you add from the ceiling fan to the light switch and fan switch should be a standard two-conductor electrical cable with #12 or #14 guage wire that has a black and a white wire. Connect them as shown in *Figure 7-9b*. You will end up with an unused white wire. When the wiring is complete, reinspect the connections, then turn on the circuit breaker. Everything should work properly. If there is a problem, pull the chain switches so the fan is off and the light is on. *Remember, the circuit is still live so proceed carefully*. Test for 115VAC with your multitester as an ac voltmeter at the circuit points indicated in *Figure 7-9*. First, measure to make sure there is 115VAC at the black and white supply wires of the branch circuit. Then test for 115VAC at the light wire connector connections, and at the wire connector connections in the outlet box for the blue and black leads for the fan light and fan, respectively. If any voltage is not present, throw the circuit breaker and recheck all the electrical connections. Using your multitester and the schematic of *Figure 7-9b*, you will solve your problem.

SUMMARY

Common electrical circuits around the home provide excellent cases where you can use your multitester for troubleshooting a problem. In this chapter, we have reviewed a number of examples. The next chapter will discuss automotive measurements.

Figure 7-9. Ceiling Fan Wiring

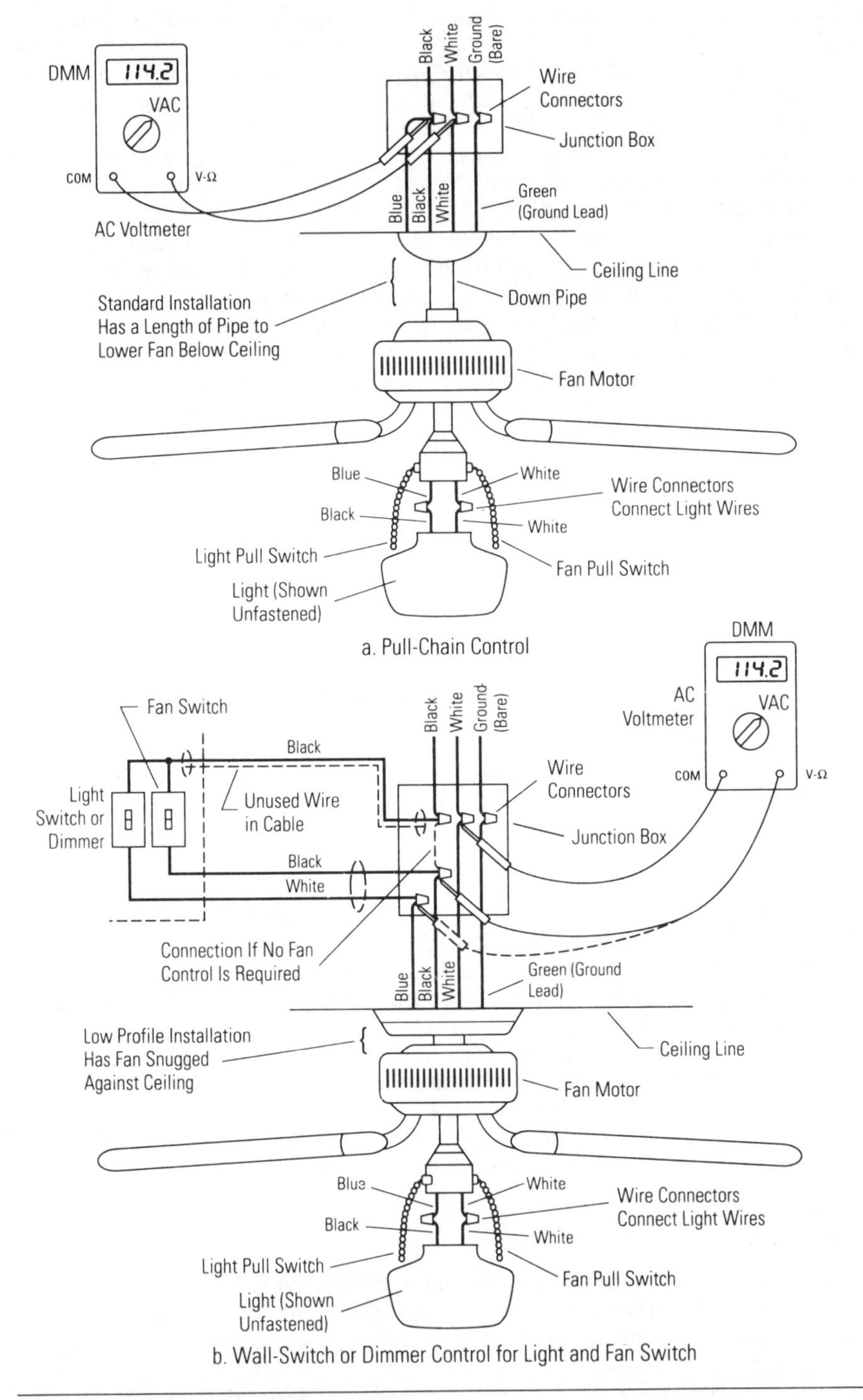

CHAPTER 8

Automotive Measurements

An older automotive electrical system consisted of a lead-acid battery, alternator, starter motor, ignition switch, and various lighting circuits. A modern system includes a variety of electrical and electronic devices, used not only in the basic electrical system, but also in almost every subsystem of the vehicle. Gone are the days when the automobile mechanic could troubleshoot anything on the car with a test lamp. To troubleshoot successfully modern automotive systems, the automotive technician must know how and what measurements to make, and be able to make them accurately and precisely. This requires a good understanding of how to correctly use multitesters. The purpose of this chapter is to present some basic automotive troubleshooting techniques using multitesters. With these techniques and a basic understanding of electrical principles such as voltage, current and resistance, you will be able to recognize the operation of several automotive systems and successfully troubleshoot them.

BATTERY, ALTERNATOR AND REGULATOR

The battery may be considered the heart of an automobile electrical system. Every electrical and electronic device on a vehicle draws its power from the battery. The battery, however, cannot perform as a continuous reservoir of electrical power unless the charging system keeps it operating at full capacity.

An automotive charging system is shown in *Figure 8-1.* The alternator is the principal part of the charging system. When the vehicle engine is running, the alternator generates a voltage and delivers it both to the battery and to the rest of the electrical system. The charging system contains a regulator, which can be either internal or external, that acts as an automatic control in the system. Without a regulator, an alternator would output its highest possible voltage continuously. Such a high voltage would exceed the limits of many components in the electrical system, including the battery and the alternator itself, and would result in extensive system damage. Let's look at some charging system measurements that can be made using a multitester.

Battery Measurements

It is all too common that the automobile battery is ignored until it fails to crank the engine. And as though this inconvenience is not enough, lack of proper maintenance will considerably shorten the useful life of a battery. Most car batteries today are of the type called "maintenance free." With an older-type battery, the status of the battery's charge was checked by testing the electrolyte with a hydrometer. With a maintenance-free battery, you test the charge

Figure 8-1. Automobile Charging System

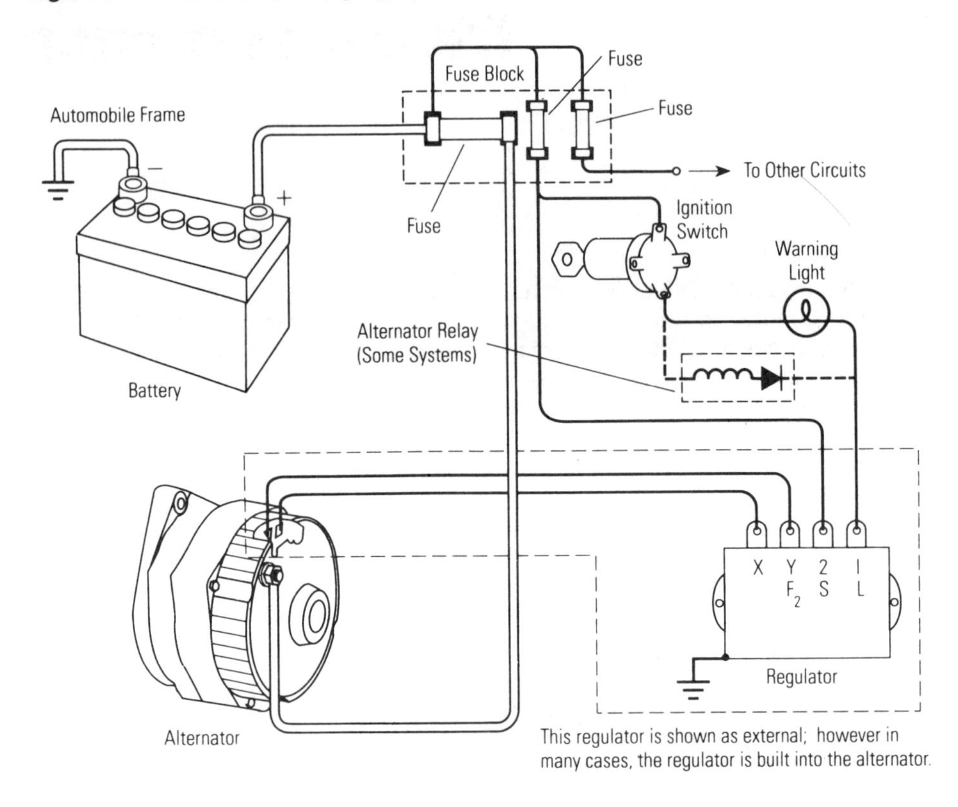

by measuring the no-load voltage across the battery terminals. *Figure 8-2c* shows the measurement with a DMM.

First, remove any current drain from the battery by disconnecting the cable from the positive battery terminal. (If the terminal and cable are corroded, clean them before proceeding according to the section on Battery Terminal Corrosion.) Measure the voltage as shown in *Figure 8-2,* and compare the voltmeter reading to the no-load test voltage in the battery charge state table *(Figure 8-2b).* This voltage test tells only the battery's charge state, not the overall condition. The overall condition is best determined under load.

WARNING

When working in the engine compartment, keep the test leads and your hands away from moving parts such as the fan and belts.

A load test that can be done at home without a special setup is also shown in *Figure 8-2c.* The battery terminal voltage is measured with a multitester while the current drawn by the starter is measured with a Hall-effect clamp-on ammeter — a clamp-on multitester that measures dc current.

Make sure the cable to the positive battery terminal is properly connected. Place the voltmeter test leads across the battery connectors, and the clamp-on ammeter around the large cable running from the positive terminal of

Figure 8-2. A No-Load and Simple Load Test for a Battery

LOAD TEST As Engine Is Cranked (Cold)		
Type of Car	**Load Current**	**Battery Voltage**
Ford 8 Cyl.	758A	10.6V
Ford 8 Cyl.	498A	10.8V
* Chrysler 8 Cyl.	490A	9.2V
Toyota 8 Cyl.	489A	11.2V
Toyota 8 Cyl.	468A	11.4V

*Old Battery, Slow Starting

a. Current and Voltage While Starting

NO-LOAD TEST	
Voltage	**% Change**
12.6V	100%
12.45V	75%
12.3V	50%
12.15V	25%

b. Battery Charge State

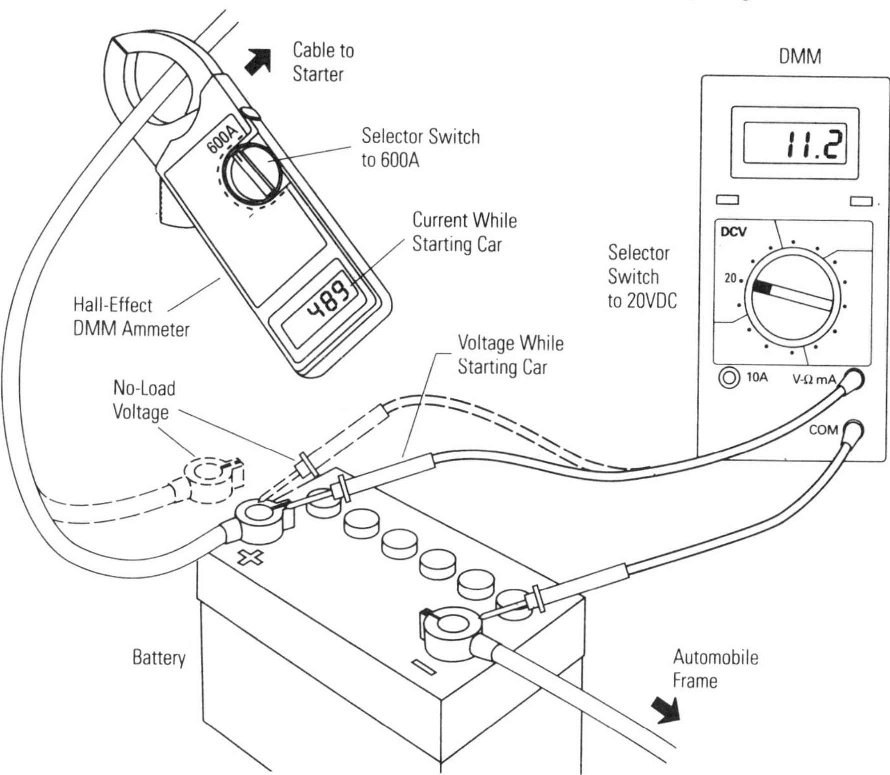

c. Actual Measurement

the battery to the starter. Read the voltmeter and the ammeter while the ignition switch is held closed to crank the engine for starting. The current reading will drop sharply after the car starts and the starting switch is released. Note that the current range for the ammeter must be from 600A to 1000A. The voltage range is 20VDC. Five typical readings are shown in *Figure 8-2a.* If the voltage reading falls below 7.2V, the battery should be charged or replaced.

Battery Terminal Corrosion

One of the most common problems with automotive electrical systems is battery terminal corrosion. If there is a high-resistance corrosion built up between the cable connector and the terminal, there will be a significant voltage drop at the cable connectors. Therefore, when measuring battery voltage under load, as in *Figure 8-2*, be sure that the multitester test lead probes are on the battery cable connectors and not on the battery terminals so the voltage reading will reveal if there is a drop. If the terminals are corroded, apply a thick solution of baking soda and water on the connectors and terminals and let it set for awhile. Rinse thoroughly with water to remove all of the solution and corrosion products. Then clean the cable connectors and terminals thoroughly by wire brushing. Scrape the inside of the connectors and the outside of the terminals with a knife or file until they are bright and shiny. When everything is clean, apply a thick coat of GB OX-GARD™, an anti-oxidant compound to the terminals and connectors, tighten the connectors properly (be careful not to overtighten), and then recheck the battery voltage under load to make sure the voltage drop has been eliminated.

LOCATING CURRENT DRAINS

In Chapter 2, we discussed how to locate and measure the current drain from the battery by all the accessories. Underhood lights, glove-box lights, trunk lights and dome lights are candidates for drawing excess current if they are stuck on. Check them carefully if excess current is indicated.

ALTERNATOR MEASUREMENTS

The exploded view of *Figure 8-3* shows that an alternator consists of a rotor inside a stator enclosed within a housing. The alternator changes mechanical energy into electrical energy. The mechanical energy is supplied by the engine through a drive belt to spin the rotor inside the alternator. The rotor spinning inside the stator produces a rotating magnetic field. The stator is a stationary set of windings attached to the housing. The rotating magnetic field of the rotor generates an electromotive force in the stator windings which provides an ac output voltage from the alternator.

The alternating output voltage produces an alternating current in any circuits connected to the alternator output. An automobile electrical system is designed to use direct current, so the ac voltage must be rectified (changed) into dc before it can be used by the automobile's electrical system. This is

Figure 8-3. Exploded View of Alternator

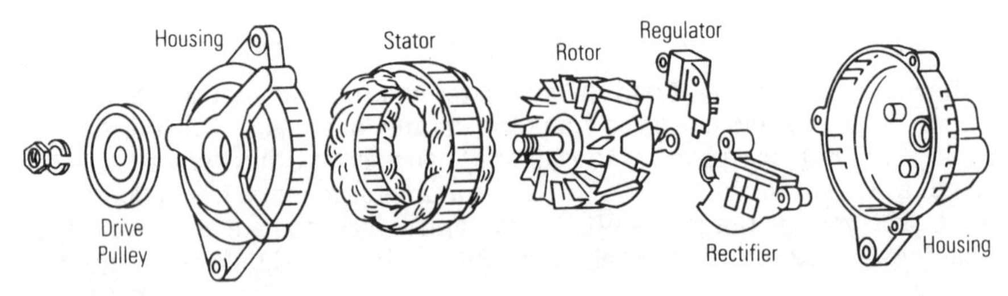

™OX-GARD is a trademark of GB Electrical Inc.

accomplished with diodes connected in an arrangement which is called a rectifier circuit. A rectifier and regulator, packaged inside the alternator, are shown in *Figure 8-3*. The alternator of *Figure 8-4* has such assemblies.

Alternator Diodes

If alternator diodes are defective, they can be either open or shorted. Diodes that have had excessive voltage applied to them usually short when they fail. Diodes that have had excessive current through them usually open. About the

Figure 8-4. Measurements on Diodes with Alternator Still Mounted in Vehicle

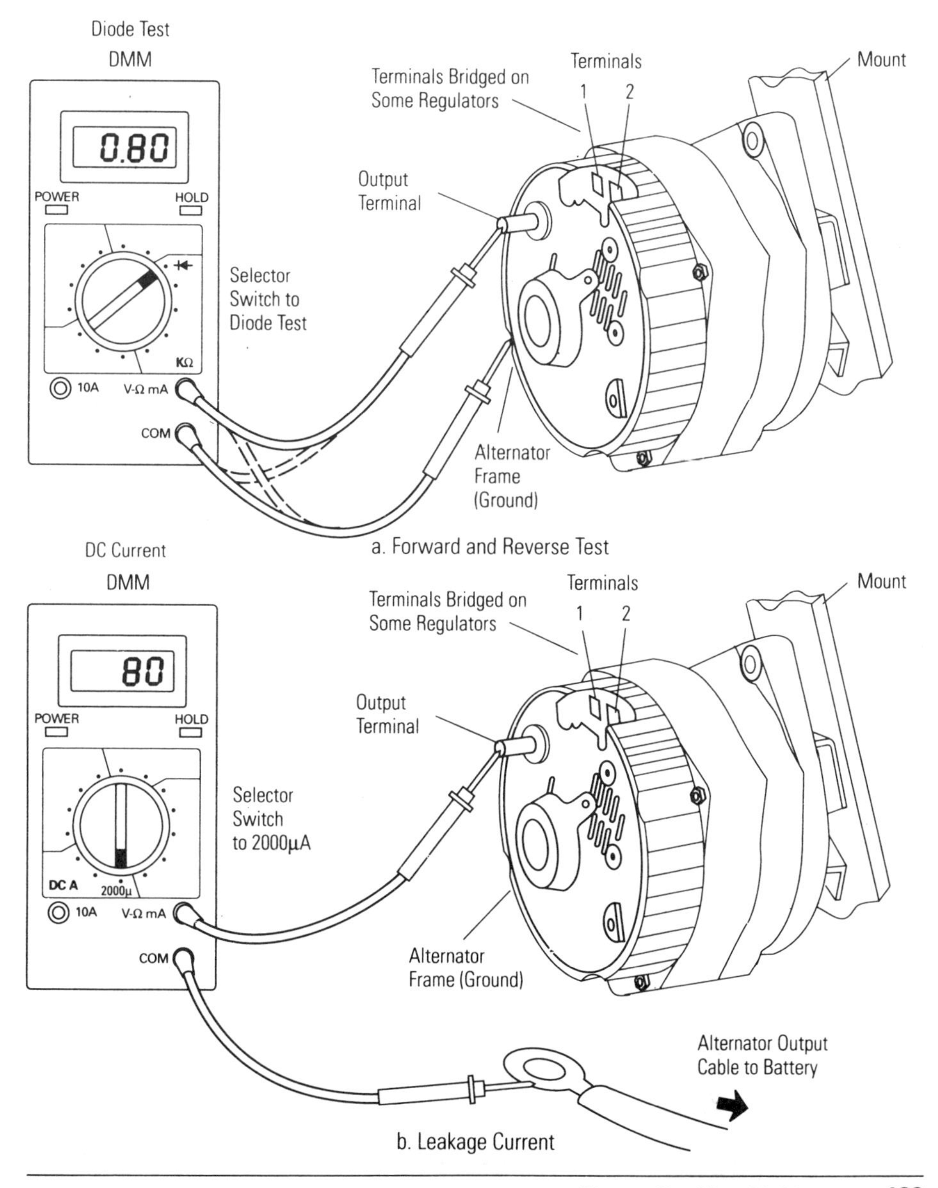

only way to find an open diode is to disassemble the alternator and individually test each diode in the rectifier assembly. Shorted diodes, on the other hand, can be found, as shown in *Figure 8-4,* by using the Diode Test function of a DMM without removing the alternator from the vehicle.

Using the Diode Test Function

With the DMM FUNCTION/RANGE selector switch set to the diode-test position (⇥), insert the test leads in the V-Ω-mA and COM jacks. With the vehicle engine **not** running, remove the connecting wires from the alternator, then, as shown in *Figure 8-4a,* touch one test probe to the alternator output terminal and the other test probe to the alternator housing and note the reading. Reverse the test probes and again note the reading. On one direction of the test, the DMM should display a "1", which indicates an open (reverse-biased diodes). In the other direction, the DMM should display 0.8V, which indicates the forward voltage of two diodes in series. If the display is about 0.4V, then one diode is probably shorted. A reading of 0.1V or less indicates two shorted diodes.

Leakage Test

If the rectifier diodes test OK, you can get an idea of how good they are by making a leakage current measurement. This time, set the DMM FUNCTION/ RANGE switch so the meter is a milliammeter (2000μA range) as shown in *Figure 8-4b.* Place the test leads in the V-Ω-mA and COM jacks. Make sure the engine is **not** running. Connect the milliammeter in series with the alternator output cable as shown in *Figure 8-4b.* The leakage current should be less than 1mA, and will probably be less than 0.1mA (80μA shown). If the leakage current approaches 20mA, the rectifier assembly should be replaced.

Voltage Regulator

A typical alternator circuit with an electronic regulator is shown in *Figure 8-5.* The description of the regulator action admittedly is simplified, but it should serve our purpose. The alternator output cable to the battery is connected to terminal 3. Current in this cable charges the battery and supplies current to the vehicle electrical system. The battery voltage must be sensed by the regulator to determine its value and tell the regulator to correct the voltage value if necessary. The battery voltage is brought to the regulator sensing circuit on terminal 2. If the voltage sensed is too low, the regulator causes a greater current through the rotor field coil, which increases the alternator output voltage. If the voltage sensed is too high, the regulator causes a lesser current through the rotor field coil, which reduces the alternator output voltage.

Transistor Q1 in *Figure 8-5* is the component that controls the current through the field coil. Delco™ alternators used on GM vehicles have a convenient "D" hole (see *Figure 8-5*) to determine if Q1 has failed or is about to fail. For the regulator to work properly, the voltage from collector to emitter must be low. As shown in *Figure 8-5,* a voltmeter, either analog or digital, can be used to measure the voltage from the collector of Q1 to ground by inserting the positive test lead probe into the "D" hole and connecting the common test lead to ground with an alligator clip supplied with the meter.

™Delco is a trademark of Delco Division of General Motors Corporation.

Figure 8-5. Alternator Electrical Circuit with Internal Regulator

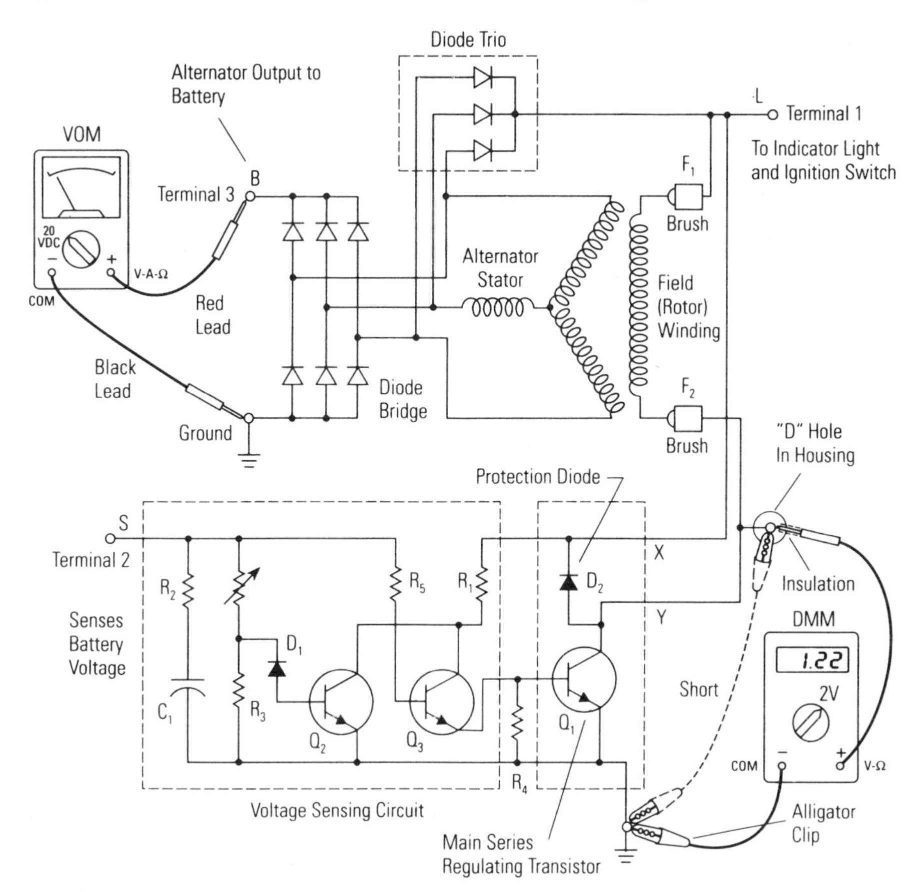

Start the engine and run it at fast idle. Fully load the alternator by turning on the headlights, radio, heater or air conditioner, and the windshield wipers. Read the voltage. If the voltage, called V_{CEsat}, is greater than 1.75V, Delco recommends that the regulator be replaced.

Checking Alternator When Internal Regulator is Defective

If the voltage at the "D" hole is close to the battery voltage, it means that the regulator is not working and needs to be replaced. The alternator warning light is on, but the question remains: Is the alternator OK? We can answer that with a measurement also shown in *Figure 8-5.* It is assumed that Q_1 is not shorted.

Start the engine and run it at fast idle. With the internal regulator defective, the alternator is not charging and the battery voltage is probably below 12V. As shown in *Figure 8-5,* with a voltmeter on the 20VDC scale, measure the alternator output at terminal 3. At the same time, insert a piece of wire in the "D" hole and short it to the alternator case (a spare multitester test lead is a good thing to use). If the alternator is good, the output voltage should immediately increase when the "D" hole connection is shorted to ground.

Checking Alternator When External Regulator is Defective

A similar measurement can be made when an external regulator is not operating properly. There are really two measurements shown in *Figures 8-6a* and *8-6b* because there are two types of regulators. The type A regulator in *Figure 8-6a* connects the field coil to ground through the regulator. Shorting from terminal F_2 to ground shorts out the regulator, and, if the alternator is good, should produce the same increased alternator output voltage as the measurement in *Figure 8-5*.

Figure 8-6. Testing for Good Alternator when External Regulator is not Working

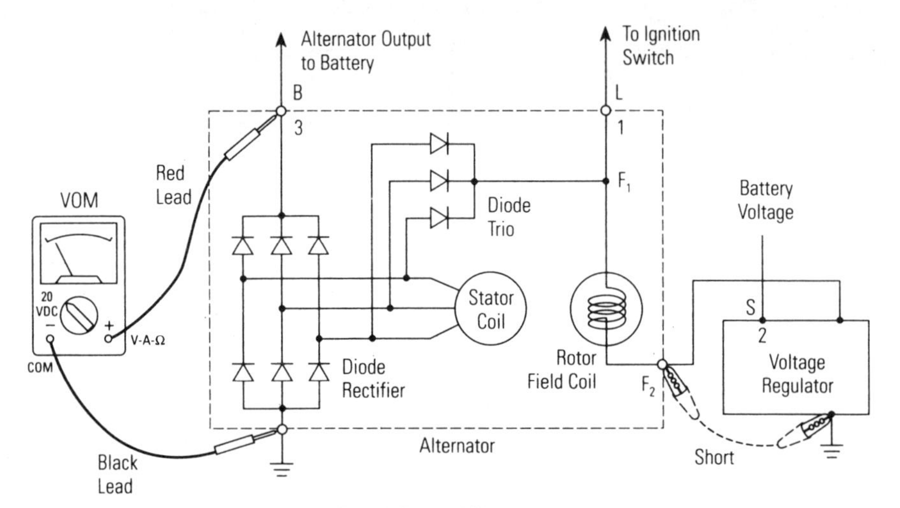

a. Type A External Regulator

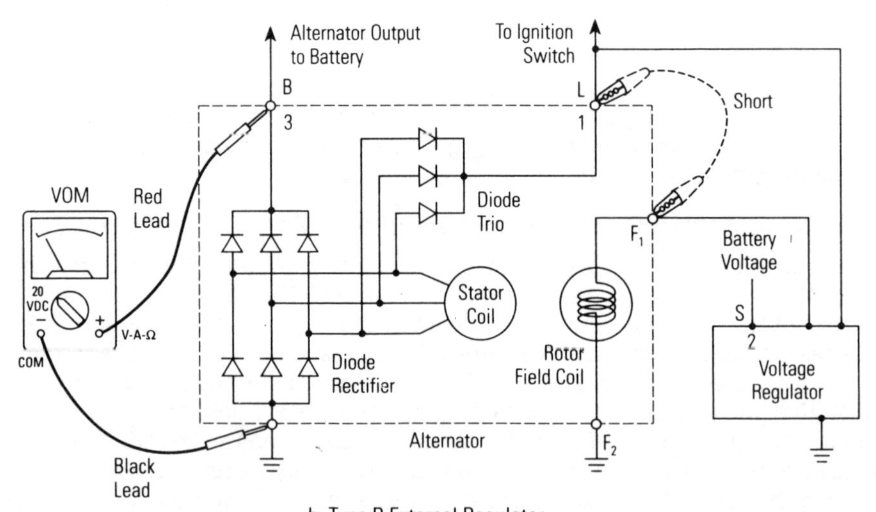

b. Type B External Regulator

The type B regulator in *Figure 8-6b* connects between terminals 1 and F_1. Terminal 1 is the battery voltage from the ignition switch and terminal F_1 is the ungrounded end of the field coil. A short between terminals 1 and F_1 should produce the same results, if the alternator is good, as the measurement in *Figure 8-5.*

Measuring the Overall Alternator

One way to measure the overall performance of an alternator is to measure the output current, output voltage, and field current all at the same time. *Figure 8-7* shows such a measurement using a DMM for the output voltage, and Hall-effect ac/dc current clamp-on multitesters for the field current and the output current. Hall-effect clamp-on ammeters are used because dc current must be measured. The 200A ranges are used on the clamp-on meters, so care must be taken not to measure starting currents because the output current can run as high as 800A. *Do not put clamp-on meters on wires until engine is running.*

Figure 8-7. Overall Alternator Performance

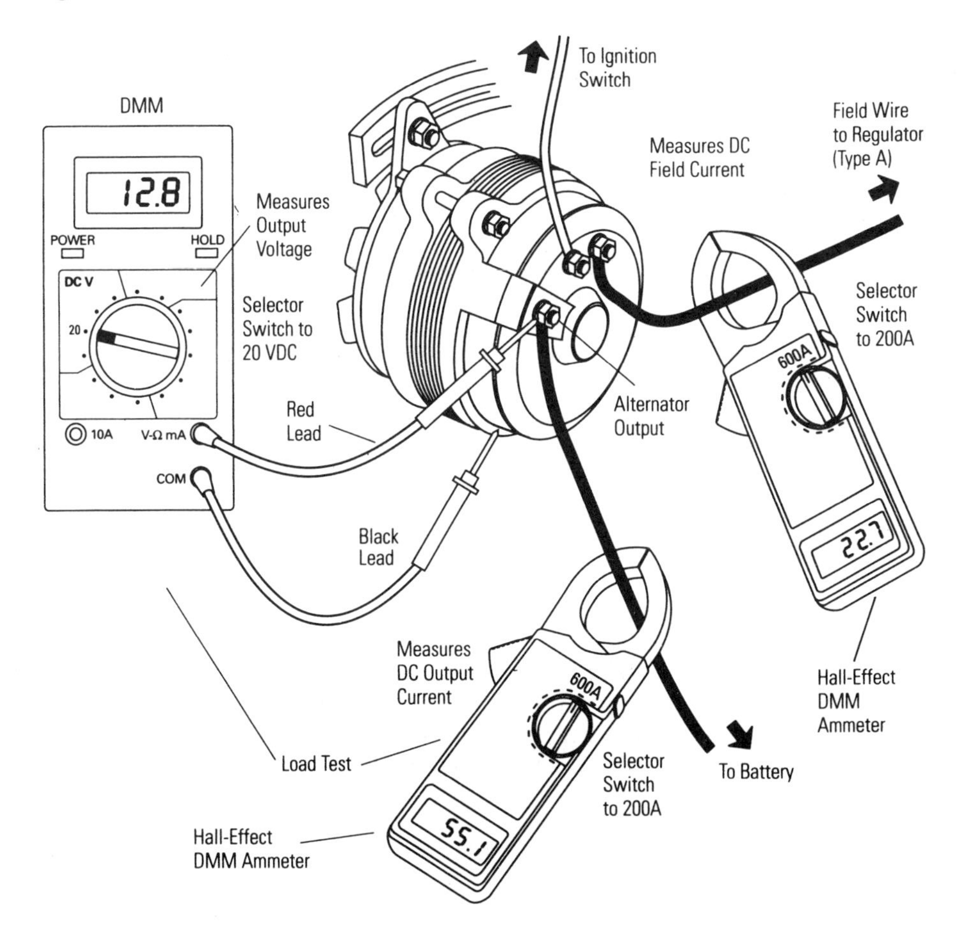

The nominal operating currents and voltages should be on the name plate on the alternator that gives the model number and operating conditions. If not, an automobile dealer's parts department or a parts house should have the data.

Testing a Disassembled Alternator

If you find that your alternator is bad, it must be disassembled to test it further. If you decide to do this, note carefully the arrangement of parts as you disassemble it so you do not omit one of the insulating washers or sleeves when you reassemble it. Use *Figure 8-3* as a guide.

The respective cathodes and anodes of the diodes are electrically connected to the stator windings as shown in *Figure 8-5*. They are mounted and connected in the alternator case or a separate mechanical assembly (heat sink) that will dissipate the heat that the diodes generate.

In the alternator of some later model cars, as shown in the schematic of *Figure 8-5*, the diodes are in the form of a diode trio for the field current and a rectifier bridge for the stator windings output. Representative disassembled parts are shown in *Figure 8-8*. Note that some of the diode heat-sink assembly is insulated. The diode trio is measured as shown in *Figure 8-8a*, and the rectifier bridge is measured as shown in *Figure 8-8b*. The readings for an ohmmeter and diode-check are shown in the table. The rotor and stator can be measured for opens or shorts to the case by using a VOM or DMM as an ohmmeter.

ENGINE IGNITION

One Cylinder Systems

An ignition system for a one cylinder engine is shown in *Figure 8-9a*. Its electrical circuit consists of the battery, ignition switch, ignition coil, distributor, and a spark plug in a combustion chamber. Early systems used a set of contacts, called points, in the primary circuit of the ignition coil. The points close and open again quickly only at one point in the rotation of the distributor, which is mechanically coupled to the crankshaft. The momentary closing of the points causes a pulse of current in the primary, which, through magnetic coupling, causes a very high-voltage pulse in the secondary. The secondary has a very large number of turns compared to the primary. The high-voltage pulse causes the spark plug to arc and ignite the air-fuel mixture compressed in the combustion chamber. The burning of the air-fuel mixture in the combustion chamber expands the gas in the chamber. The expanding gas pushes the piston down and rotates the crankshaft. Some external cranking is required to get the system started.

Because the points burned due to the arcing across them, the points have been replaced with an electronic component — in *Figure 8-9a*, it is a transistor. A magnet on the distributor shaft sweeps by a sensing coil at a particular point in the distributor shaft rotation. The pulse produced in the sensing coil triggers the electronic control circuit and turns on the transistor. Because the transistor acts as a switch, it does the same thing as closing the points to cause the spark plug to ignite the air-fuel mixture.

Figure 8-8. Measuring Disassembled Alternator Diodes

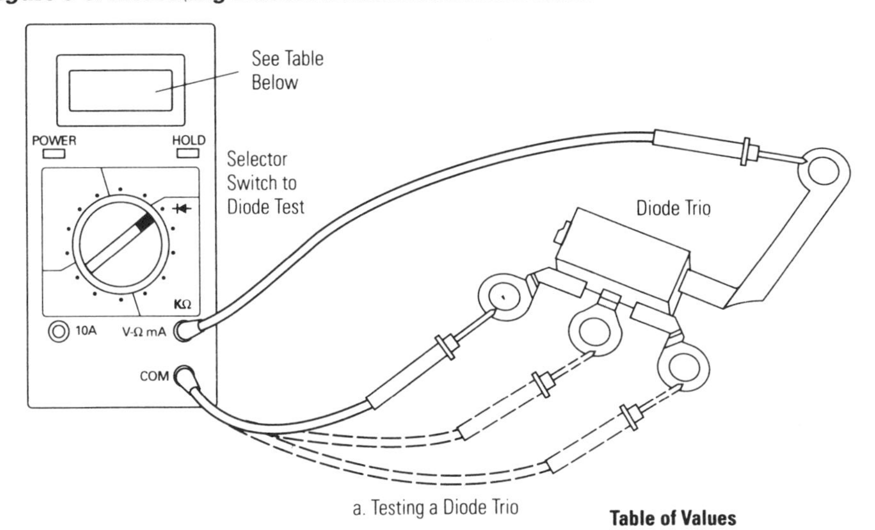

a. Testing a Diode Trio

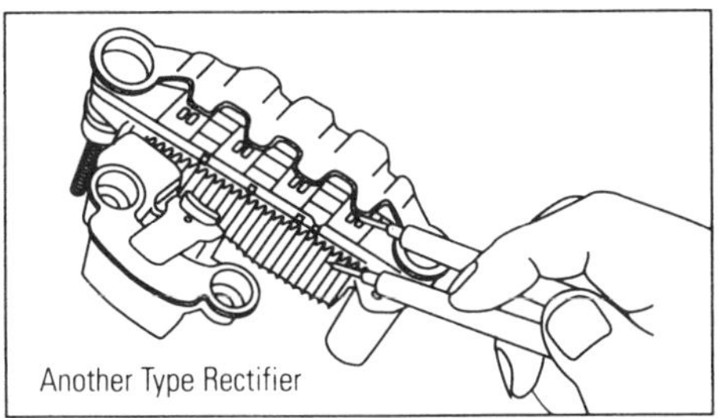

Table of Values

DMM or VOM as an Ohmmeter or DMM on Diode Check			
Meas.	Ohmmeter		Diode Check
	VOM	DMM	DMM
Reverse	∞	*1.000	*1.000
Forward	45Ω	45Ω	0.6 – 0.7V

*1 may flash and zeroes may not be present on some DMMs

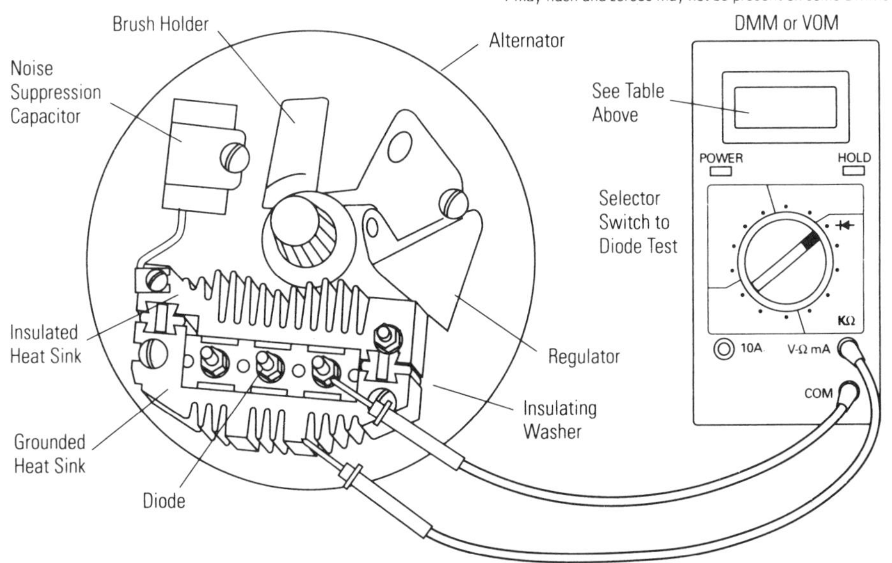

b. Testing a Rectifier Bridge

Figure 8-9. Engine Ignition

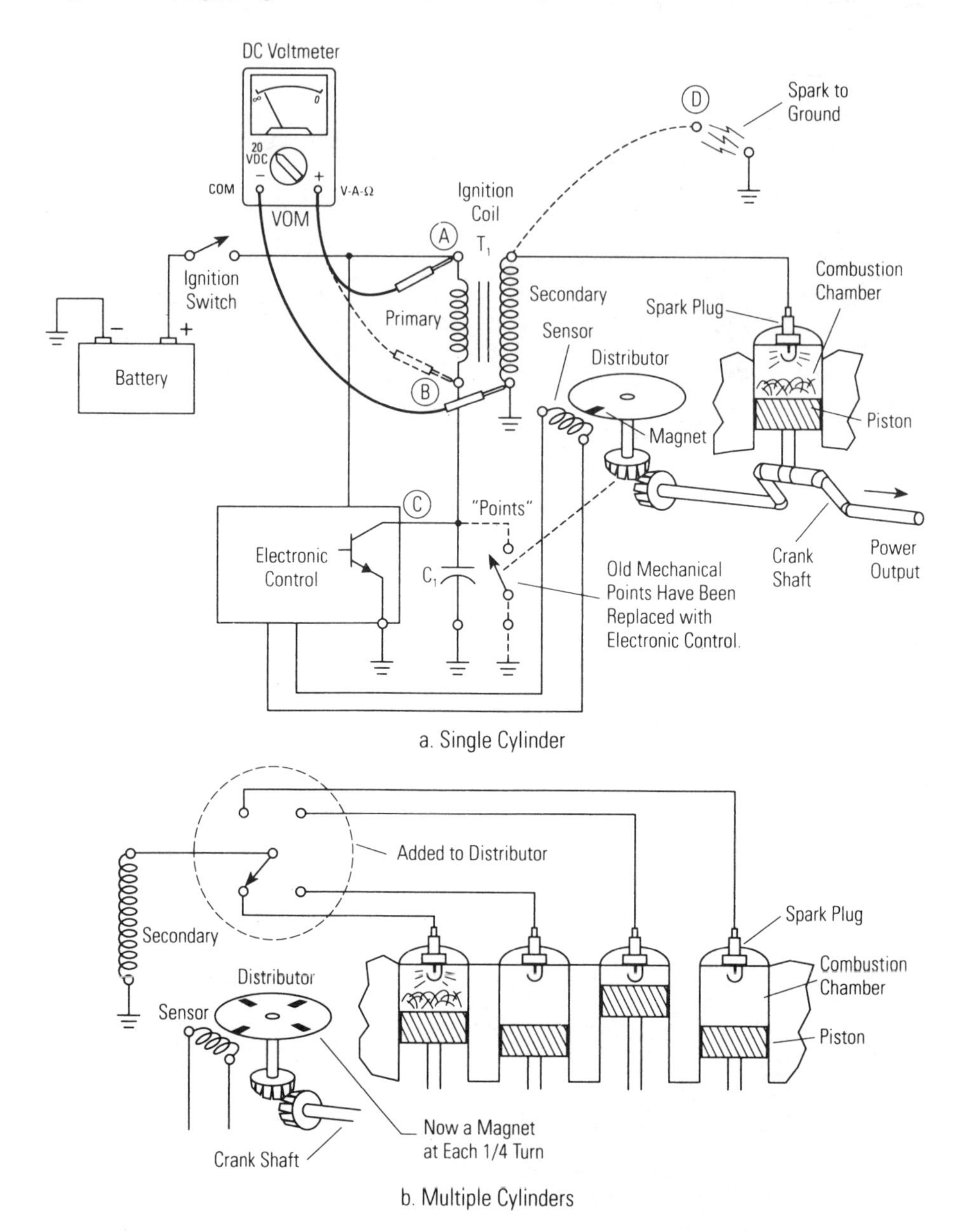

a. Single Cylinder

b. Multiple Cylinders

Multiple-Cylinder Engine

Multiple-cylinder engines have the same basic circuit, as shown in *Figure 8-9b*, except that the distributor is different. Another part has been added to distribute the high-voltage pulse to the correct cylinder, and more magnets have been added corresponding to the number of cylinders. The magnets are spaced around the distributor to provide the proper high-voltage at each spark plug at the correct time. The action is the same as for one cylinder, except it occurs for each cylinder in a timed sequence.

Lawn Mower Systems

A lawnmower engine's ignition system (single-cylinder) follows the same basic principles, and may be either of the magneto type or of the solid-state type shown in *Figure 8-10*. Both types have a powerful magnet mounted on a flywheel attached directly to the engine crankshaft. Both have a coil assembly mounted close enough to the flywheel so the magnetic field from the magnet is coupled to the coil assembly. The magneto system has a set of points which operate basically the same as the one-cylinder system. The opening of the points generate the high-voltage spark for the plug through a secondary on the coil assembly. Twisted or bent engine shafts that throw off the mechanical timing of the points is a major problem in the magneto system.

The solid-state system coil assembly is a bit different. It contains a charge coil, a trigger coil and the high-voltage secondary pulse transformer. The electronic switch is an SCR (silicon controlled rectifier.) The basic operation is as follows: The magnet on the flywheel, sweeping by the coil assembly, charges the capacitor, C_1. A pulse from the magnet in the trigger coil turns on the SCR, which discharges C_1 through the coil assembly and produces 25,000 volts in the secondary pulse transformer to cause the spark in the plug.

Ignition System Measurements

CAUTION

Do not attempt to use an ordinary multitester to measure the high voltage at the ignition coil secondary or at a spark plug. The multitester could be severely damaged.

Figure 8-10. Lawn Mower Solid-State Ignition
(Courtesy of Sears, Roebuck and Co.)

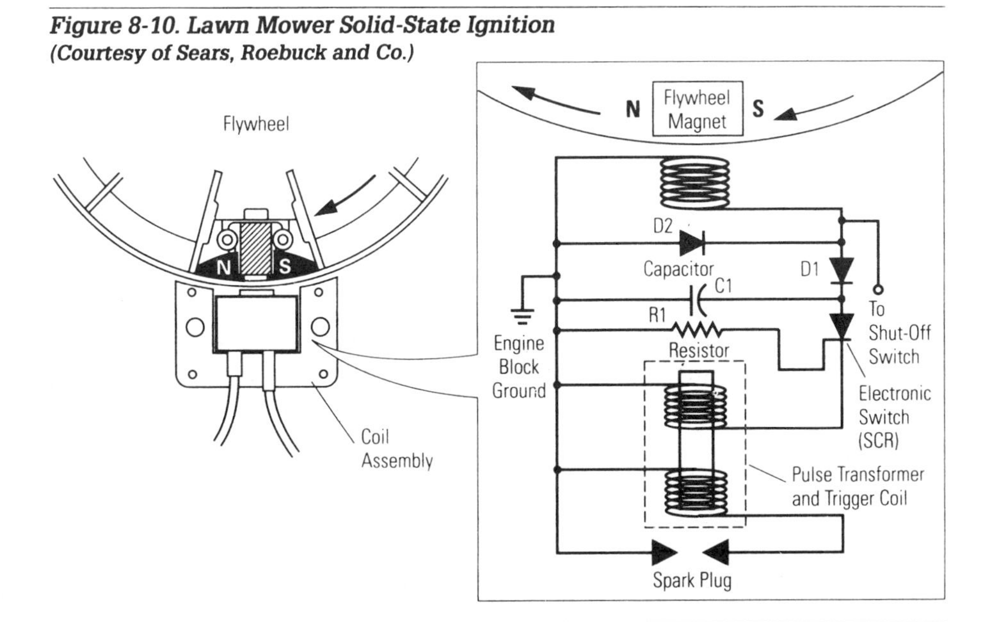

If you suspect that an engine's ignition system is not operating properly, the first thing to do is measure with a voltmeter from point A to ground as shown in *Figure 8-9. With the ignition switch on but the engine not started,* you should read battery voltage at A (across the primary). If A does not have battery voltage, check the connections to the primary of the ignition coil and check for corrosion at the battery terminals. If A has battery voltage, check point B as you crank the engine. You should see a pulsing or flicking of the needle or jumping of a DMM meter reading if the electronic control is working properly. If B does not pulse as the engine is cranked, remove the electronic control module and test from the output, C, to ground with an ohmmeter or with a diode checker. If there is a short, the electronic control module must be replaced. If there is an open circuit, see if you can isolate the output transistor to measure its junctions and determine if one or both of the junctions are open.

If there is proper operation at point A and B, turn off the ignition switch and remove the spark plug wire from the spark plug. Arrange the wire at D to lay on the engine. *Do not hold onto the spark plug wire!* Now crank the engine. If a spark jumps from D to ground, the system is operating properly, and the trouble is probably a bad spark plug. If the system is a multiple cylinder system, check each plug wire this way. If there are problems, check the distributor rotor or the contacts inside the distributor cap that connect to the spark plug wire terminals. They may be burned or corroded. Burnish the contacts or buy a new distributor rotor and cap.

If there is no spark to ground at D, turn off the engine, remove the ignition coil and measure the coil primary and secondary for opens or for shorts to the case. Sometimes the measurements must be made when the coil is hot to detect a defective condition. Spark plug wires could also be open. This is a common problem. Check each one with an ohmmeter, but be aware that they are a high-resistance wire, like 10,000 ohms per foot. They should not be anymore than 50,000 ohms. Gently twist the wires as you make the measurements to detect intermittents. Replace any wires that are defective.

Capacitors

Automotive capacitors (condensers) can be checked with a multitester the same as discussed in Chapter 4. This is one test where the analog meter does the job better than a DMM. *Before any measurement is taken, short across the capacitor terminals to discharge the capacitor.*

Distributors

In today's automotive vehicles, there also may be distributors with Hall-effect pickups or with optical pickups. All put out pulses to trigger the control module to produce a high voltage to the spark plugs. If you suspect the distributor is the problem, go to a qualified service outlet and have the distributor checked because of minimal checks that can be made with a VOM or DMM.

SUMMARY

In this chapter we have shown some common measurements that can be made to troubleshoot problems in an automotive electrical system. VOMs and DMMs have been used as voltmeters, ammeters and ohmmeters in a variety of applications. The next chapter will deal with measurements on the circuits contained in power tools found around the workshop.

CHAPTER 9

Tool Control Circuit Measurements

In this chapter, we will examine the circuit operation of several power tools and the control circuits that are used for them. We will also look at several measurements that you can make on these circuits with your VOM or DMM, as long as the power tool is *not* of the double-insulated type.

WARNING

The assembly or reassembly of double insulated type of tools requires checking with special high-voltage testing equipment by trained and qualified technicians. These tools should not be disassembled by the user for testing or parts replacement. *A severe electrical shock hazard may occur if the replacement parts are not properly installed and then tested.* Service should be performed by a qualified repair center.

Ground Test

If your electric drill motor is not of the double-insulated type, but instead it has a metal case and a three-prong power plug, you should be sure that the case is grounded, and, in addition, be sure that there are no bare wires shorted to the case. Either condition is potentially dangerous.

The measurements to be made with an ohmmeter are shown in *Figure 9-1.* If everything is normal, the wires that carry the ac power to the drill from the outlet should each measure infinity ohms to the drill case, and the ground wire should measure zero ohms (continuity) to indicate it is tied to the drill case.

Figure 9-1b shows schematically how the ground pin to case measurement is made. This is a typical schematic of most electrical hand tools. If the variable speed control is not in the circuit, terminals A and B would be shorted together. If the reversing switch is not in the circuit, the armature may be connected on either end of the field or between the two field coils, and the ON/OFF single-pole single-throw switch is just in series with the black power line.

VARIABLE-SPEED CONTROLS

There have been several methods used to control the speed of electric motors since variable-speed appliances and power tools appeared on the market a few decades ago. All of the methods use the effect that a electric motor runs slower when the operating voltage is reduced.

Figure 9-1. Measurements on Electric Hand Drill

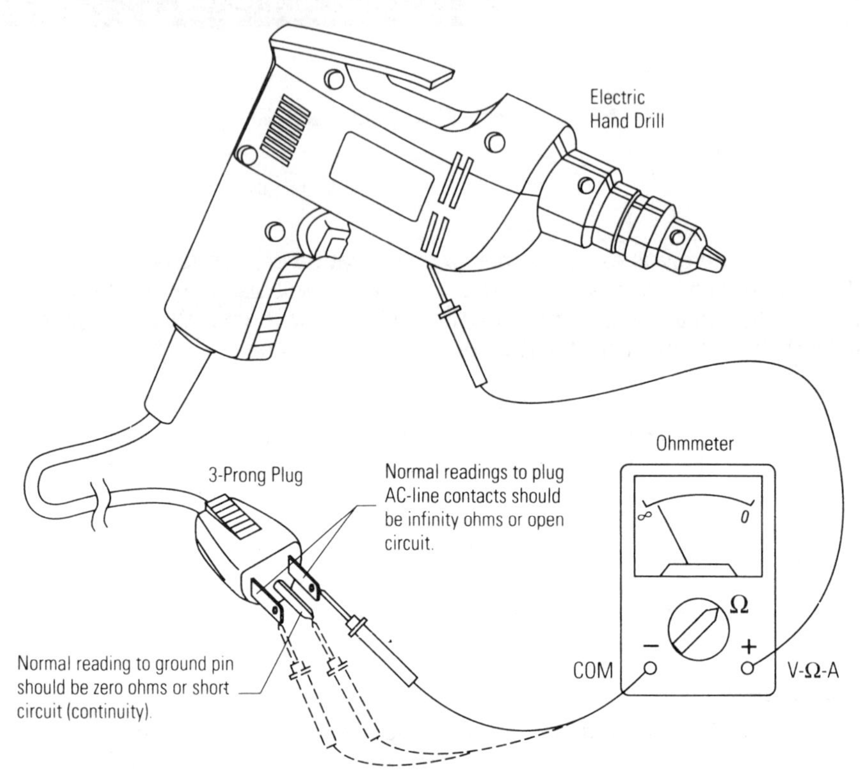

a. Measurements for Shorts Between Case and Power Cord

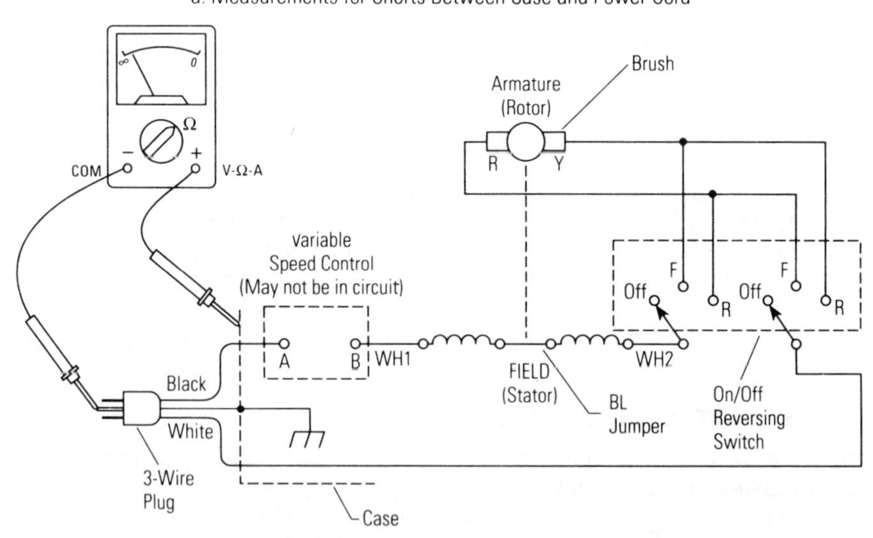

b. Schematic (Typical of Most Electrical Hand Tools)

Variable R and L Controls

Many of the early controls used a variable resistor connected in series with the motor. Adjusting the resistor changed the amount of voltage applied to the motor, and thus, changed the operating speed. This method wasted a great deal of power at slow speed because most of the line voltage was dropped across the resistor. Another type reduced the motor voltage by connecting an inductor instead of a resistor in series with the motor. This circuit was more efficient than the resistive power control, but difficult to make variable.

Solid-State Controls

A much more efficient, present-day method of speed control makes use of solid-state devices. The term *solid-state* means that an electrical circuit has diodes, transistors, thyristors, and/or integrated circuits in it. The control circuit, which may also include resistors, capacitors and inductors, provides a variable operating voltage to the motor. It works on the principle that the greater the average operating voltage, the faster the motor speed.

Diode Speed Control

The simplest solid-state speed control is shown in *Figure 9-2.* It either switches a diode rectifier in series with the motor for the low speed, or bypasses the diode for high speed. *Figure 9-2* also shows the waveforms associated with each speed. The diode simply blocks current during one of the power line voltage alternations, effectively cutting the average operating voltage in half. This reduces the motor operating speed by about one-half without wasting any power. Switch S_1 is shown as a trigger switch on a power tool such as an electric drill. When the trigger is not pulled, S_1 is in the OFF position. By pulling the trigger half way, S_1 connects the diode in the circuit and the drill runs at low speed. When the trigger is pulled all the way, S_1 bypasses the diode, and the motor runs at full speed. Notice that the motor must be capable of running on both ac and dc voltage if this type of control is used. Such a motor is called a universal motor.

Testing the Two-Speed Diode Control

As previously stated, if the electric drill motor is not of the double-insulated type, then you can troubleshoot it with your multitester. The ways to make several common measurements on the two-speed circuit of *Figure 9-2* are shown in *Figure 9-3.* The two-speed circuit has very few parts, so testing it is relatively simple. Here are four common symptoms that could possibly happen when a two-speed hand tool does not work:

1. The motor does not run at any speed setting.
2. The motor only runs when the trigger switch is in its high-speed position.
3. The motor runs at high speed in either the low- or high-speed position of the trigger switch.
4. The motor runs only when the trigger switch is in the low-speed position.

Let's look at these symptoms one at a time.

Figure 9-2. Simplest Solid-State Speed Control

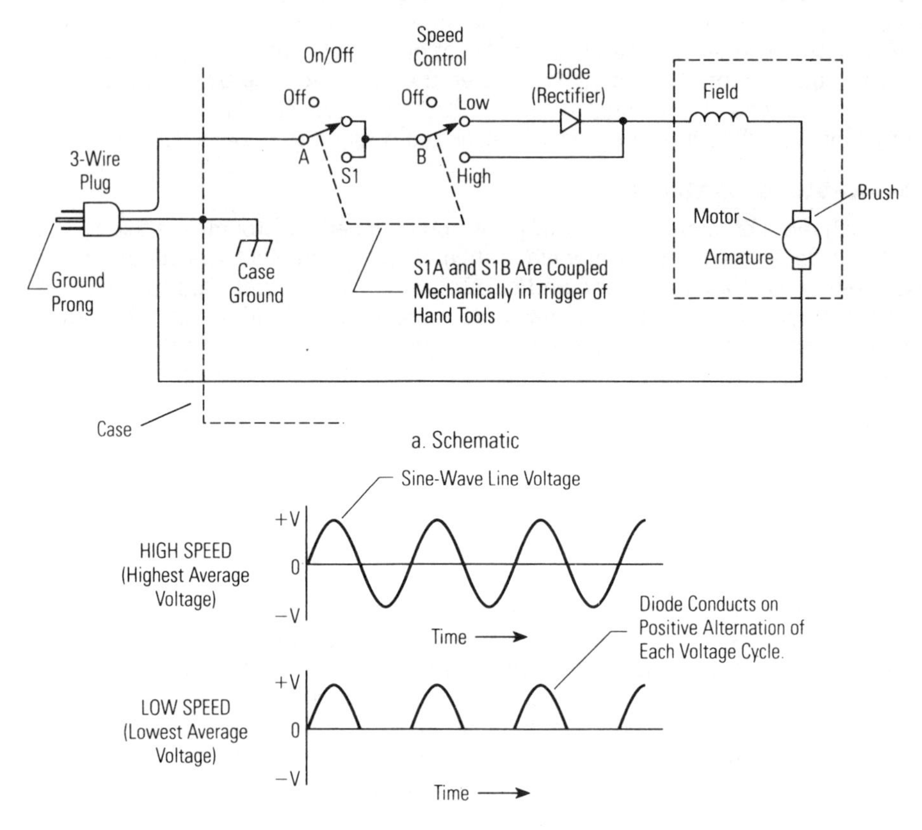

a. Schematic

b. AC Voltage Waveforms

First, if the motor does not run at any speed, this indicates an open circuit, either in the wiring to the trigger switch, in the switch itself, in the power cord, or in the motor circuit. A continuity check with an ohmmeter on OHMS or the audible tone continuity position of the FUNCTION/RANGE switch should indicate the cause of the open circuit. Specific measurements with a meter at different circuit locations are shown in *Figure 9-3.* When the trigger switch is pulled to the high-speed position, measurement M2 indicates a low resistance if the circuit is operating properly. Since there is an open circuit, M2 will read infinity ohms. Measurements M1 and M3 will indicate if the trouble is in the power cord or lines bringing power into the tool. The cord, which is the most common problem, can then be measured as shown in *Figure 9-3b.* Another common problem, worn or broken brushes, can be located with the M1 measurements. M1 also locates opens in the motor field windings. Measurements M3 and M4 locate problems in the trigger switch.

Figure 9-3. Diode and Continuity Measurements

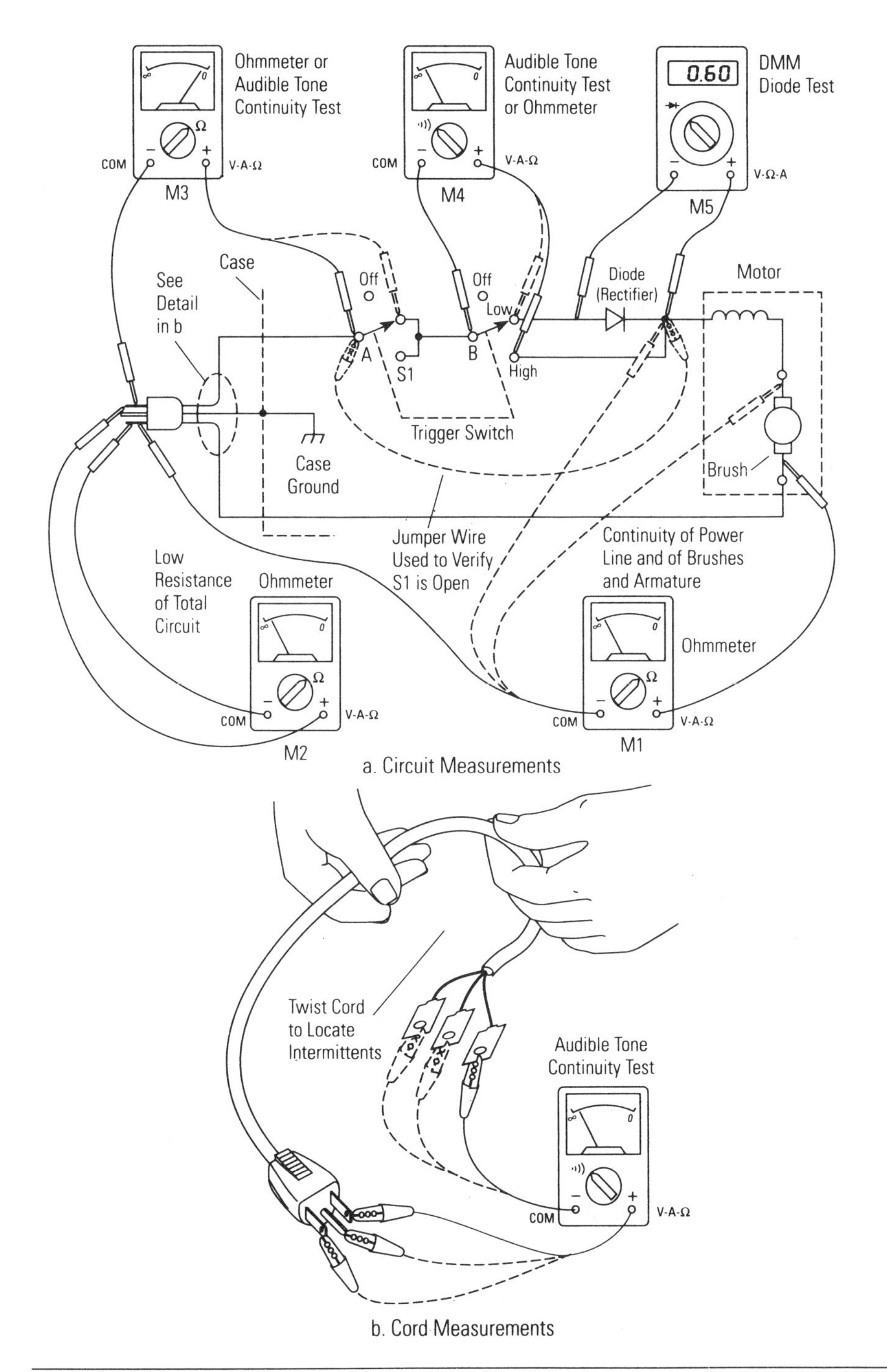

a. Circuit Measurements

b. Cord Measurements

If the control switch indicates open, you can use a jumper wire to short out the switch as shown in *Figure 9-3a*. **CAUTION:** *Be sure the power cord is unplugged from the outlet before attaching the jumper wire.* Now, when the drill's power cord, if it has been verified as good, is plugged in, the motor should run at high speed. The shorting jumper bypasses the switch and diode rectifier so the only thing left in the circuit is the motor and connecting wires. If the motor still fails to run, verify that you have power at the outlet as in Chapter 1 or 2. If there is power at the outlet, check the power cord again, and all circuit screw or wire connector connections for broken leads or bad connections.

For the cord measurement of *Figure 9-3b*, a multitester with an audible tone continuity tester can be used effectively to find an intermittent break in the power cord. Set the multitester FUNCTION/RANGE switch to the audible continuity ( )))) position. Connect the test leads to the plug prongs and twist the cord, especially near the plug and the drill motor. If the cord has intermittent breaks, they will be indicated by tone bursts from the continuity tester as the break in the wire momentarily touches.

Second, if the motor only runs with the switch in its high-speed position, an open diode or open S_1 in its low-speed position is indicated. The diode may be checked with a VOM on the ohmmeter function or a DMM with a Diode Test function as shown for the M5 measurement. If a VOM is used, be sure to measure the diode in both directions as we discussed in Chapter 4. If the VOM reads high resistance (over 1000Ω) in both directions, the diode is open. Measurements M3 and M4 are used to test S_1.

Third, if the motor runs at high speed on either the low- or high-speed position of the trigger switch, then, either the diode is shorted, or the trigger switch has a short between the two speed selection positions. If the diode is shorted, it can be tested as we described above. Remember that a shorted diode will measure low resistance in both directions on a VOM. Again, measurements M3 and M4 will locate the trouble in S_1.

Now, the fourth symptom — where the motor runs only when the trigger switch is in the low-speed position. This indicates that the diode is in the circuit and that the low-speed position is OK; however, there is an open when the trigger switch is in the high-speed position. Measurements M3 and M4 should locate the problem.

Solid-State Module Speed Control

Figure 9-4 shows a hand tool circuit where a solid-state speed control module controls the average operating voltage to the motor. The external speed adjustment, which is simply a variable resistor built into the trigger switch, controls the solid-state module which controls the motor speed. The circuitry of the solid-state module is usually encapsulated so that repair is not possible. If it is defective, the entire unit must be replaced. Problems in the module, ON/OFF switch, line cord, and trigger switch with its variable resistor can be tested with your VOM or DMM as described for *Figure 9-3*. We will look at how this is done, but first let's briefly discuss how a solid-state module speed control works.

The solid-state module speed control has at its heart a semiconductor device called a *thyristor*. It provides continuous control for smooth speed

Figure 9-4. Solid-State Variable Speed Module for Hand Tool

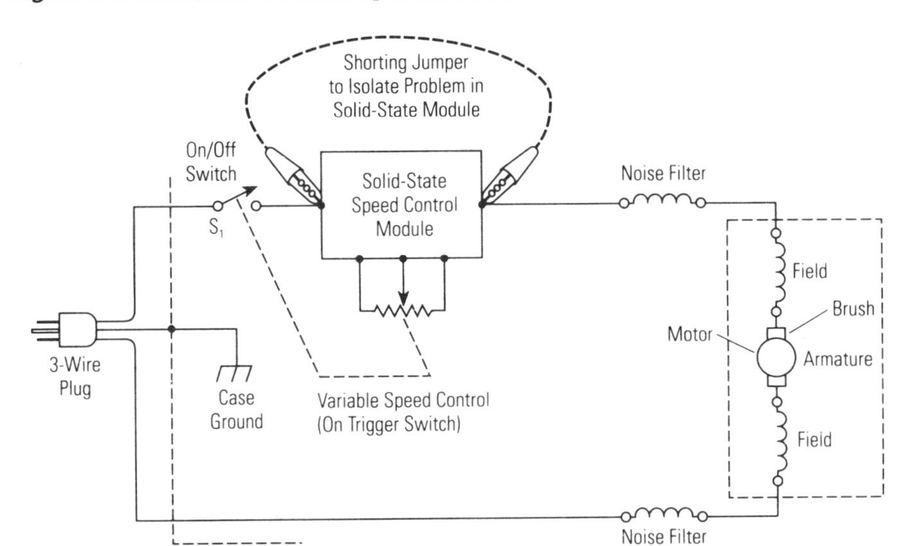

transitions. Thyristors are a family of solid-state power switching devices which include silicon controlled rectifiers (SCRs) and bidirectional triode thyristors (TRIACs). Both of these devices have a control element called a *gate*. The current to the gate determines when the device triggers from a high resistance across its power terminals to a low resistance. The device will not trigger at a particular gate current until the amount of voltage across the device reaches a certain value. In other words, at a given gate setting, the voltage across the device must increase to a certain value before the device will conduct. The gate bias may keep the device off completely and never let it conduct, or it may permit conduction to begin at any desired point on an applied sine-wave voltage cycle. Once it is turned on in a cycle alternation, it passes the full line voltage to the motor (or any other load) until the end of that half-cycle of line voltage.

A TRIAC power control circuit is shown in *Figure 9-5*. During the positive half cycle of the input sine-wave supply voltage, diode D_1 is forward biased, D_2 is reverse biased, and the gate terminal is positive with respect to A_1. During the negative half cycle, D_1 is reverse biased and D_2 is forward biased, so that the gate becomes positive with respect to A_2. Adjustment of R_1 controls the point at which conduction begins.

The variable resistor controls the gate signal and, therefore, it determines when the TRIAC turns on during each power half-cycle. When the applied voltage is a sine wave, the point at which the gate current and the voltage across the device causes the device to conduct provides an electronic time delay. The amount of the delay interval is determined by the setting of R_1, the variable resistor. The time-delay occurs within each half-cycle. The later the TRIAC is turned on, the lower the average voltage applied to the motor and the slower the motor speed.

Figure 9-5. Solid-State TRIAC Control Circuit

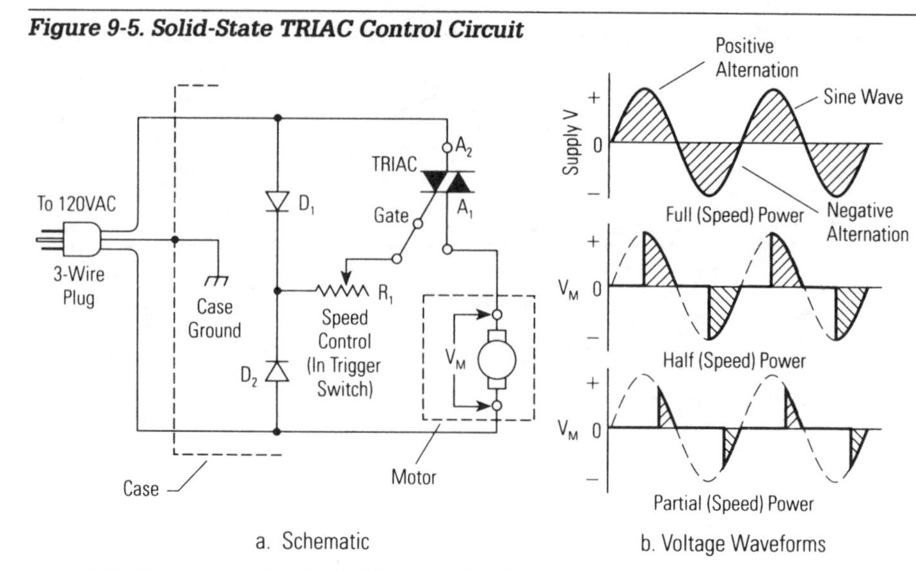

Troubleshooting the Solid-State Variable Speed Control

The power cord, the switch and the motor brushes are the most likely sources of electrical trouble on power tools, including universal electric drill motors. Of course, the field and rotor coil winding or the commutator may also cause problems. Visual inspection of the parts is, as usual, the first troubleshooting step. The measurements and techniques described in *Figure 9-3* will pinpoint the exact circuit trouble.

The electronic components included in the solid-state control circuit, if they are not encapsulated, are nearly always mounted on a printed circuit board. The same shorting jumper technique as used for the trigger switch of *Figure 9-3* can be used to short out the solid-state module and eliminate it from the circuit so that the other circuit components can be checked. This technique is shown in *Figure 9-4.* When it is determined that the module is bad, try resoldering its connections. If this does not fix it, the entire printed circuit assembly probably will have to be replaced.

TOOLS WITH ELECTRONIC READOUTS

Some radial saws, sanders, and other bench tools now have digital readouts. If trouble occurs in these tools, in most cases, troubleshooting the problem requires specialized test equipment. Usually, the readout or electronic controls are replaced as a separate unit. See your local qualified repair center for such repairs.

SUMMARY

This chapter concludes our discussions of the application of multitesters to all types of electrical measurements. Our objective was to introduce you to multitesters and show you how to use them to make a wide variety of measurements. By now, you should be comfortable with a multitester and have found it as useful as your pliers and hammer. That was our goal; we hope we have succeeded.

Appendix

Determining Resistor Values

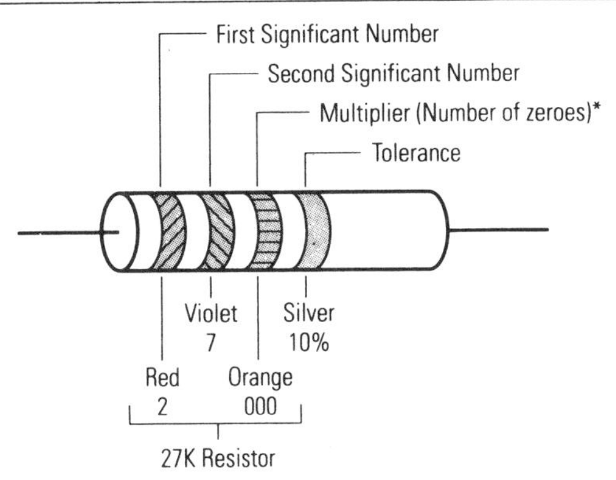

Color Bands

Color	Value
Black	0
Brown	1
Red	2
Orange	3
Yellow	4
Green	5
Blue	6
Violet	7
Gray	8
White	9
Tolerance	
Gold	5%
Silver	10%
No Band	20%

* If multiplier band is gold, the multiplier is 0.1; if it is silver, the multiplier is 0.01.

Color Code

A common way of indicating resistance values for composition resistors is to use color bands on the body of the resistor. A standard color code for the bands has been adopted by resistor manufacturers, and the numerical values they represent are shown. The first band will be nearer one end of the resistor. Read from this band toward the other end. The first band is the first significant number, the second band, the second significant number and the third band the multiplier (number of zeros that follow the first two numbers) to indicate the resistor's value. If the third band is gold or silver, this would indicate a multiplier of 0.1 or 0.01, respectively, rather than additional zeros.

As the resistors are manufactured, there will be some variation in values around the nominal desired value. The fourth band indicates the tolerance of the resistor's value from the indicated (color code) value. A gold fourth band indicates ±5%, a silver band ±10%, and no fourth band indicates a tolerance of ±20%.

The resistor shown is a 27,000 ohm or 27 kilohm resistor. It has a ±10% tolerance, so the actual value of the resistor can be any value from 24,300 ohms to 29,700 ohms. A resistor's physical size usually indicates its power rating. Resistors with less than two watts power rating weigh less than an ounce, but high-power rating resistors have large bulky bodies to dissipate the heat.

Inductance

The opposition offered by an inductor to ac is called reactance. The amount of inductive reactance is given by the following equation:

$X_L = 2\pi fL$ where
- X_L is the inductive reactance in ohms
- f is the frequency of the ac in hertz
- L is the inductance in henries.

The quantity $2\pi f$ represents the rate of change of current in radians per second. It is called angular velocity or angular frequency. The amount of inductance of a coil is measured by a unit called the henry. However, smaller units are more practical — the millihenry (1×10^{-3} henry) and microhenry (1×10^{-6} henry) are very common units in electronic circuits.

The amount of inductance in an inductor depends on the magnetic flux produced and the current in the coil. Mathematically, this may be expressed by the following equation:

$L = \frac{N\phi}{I}$ where
- L is the inductance in henries
- I is the current through the coil in amperes
- N is the number of turns of wire
- ϕ is the magnetic flux linking the turns.

Basically, all inductors are made by winding a length of conductor around a coil form made of insulating material. The center of the coil form, or core, may be empty (an air core) or filled with a solid material, which is usually magnetic. The table shows how the physical characteristics of the core and the windings around the core affect the amount of inductance produced.

Lower L	**Higher L**
Small number of turns	Large number of turns
Hard steel core	Soft iron core
Small cross-section area	Large cross-section area
Long-core, wide spacing	Short-core, narrow spacing

Glossary

Alternating current (ac): An electrical current that periodically changes in magnitude and in direction of the current.

Alternation: Either half of a cycle of alternating current. It is the time period during which the current increases from zero to its maximum value (in either direction) and decreases to zero.

Alternator (or ac generator): An electromechanical device which transforms mechanical energy into electrical energy — an alternating current. Very early users called this a dynamo.

Ammeter: An instrument for measuring ac or dc electrical current in a circuit. Unless magnetically coupled, it must be placed in the current path so the flow is through the meter.

Ammeter, Clamp-on: An instrument used to measure current by means of induction or Hall-Effect without opening the circuit.

Ammeter shunt: A low-resistance conductor that is used to increase the range of an ammeter. It is shunted (placed in parallel) across the ammeter movement and carries the majority of the current.

Ampere (A): The unit of measurement for electrical current in coulombs (6.25×10^{18} electrons) per second. One ampere results in a circuit that has one ohm resistance when one volt is applied to the circuit.

Amplification: See Gain.

Amplifier: An electrical circuit designed to increase the current, voltage, or power of an applied signal.

Analog-to-Digital Conversion or Converter (ADC or A/D): The process of converting a sampled analog signal to a digital code that represents the amplitude of the original signal sample.

Audio and audio frequency (AF): The range of frequencies normally heard by the human ear. Typically, about 20 to 20,000 Hz.

Beta (β) (h_{FE}): The current gain of a transistor when connected in a common emitter circuit.

Bias: In an electronic circuit, a voltage or current applied to an active device (transistor, diode, etc.) to set the steady-state operating point of the circuit.

Binary Coded Decimal (BCD): A binary numbering system in which any decimal digit is represented by a group of 4 bits. Each digit in a multi-digit number continues to be identified by its 4-bit group.

Binary digit (Bit): A digit in the binary number system whose value can be either 1 or 0.

Bipolar: A semiconductor device having both majority and minority carriers.

Bit: See Binary digit.

Block diagram: A system diagram which shows the relationship between the main functional units of the system represented by blocks.

Breakdown: The condition for a reverse-biased semiconductor junction when its high resistance, under the reverse bias, suddenly decreases, causing excessive current. Not necessarily destructive.

Bridge rectifier: A full-wave rectifier in which the rectifier diodes are connected in a bridge circuit to allow current to the load during both the positive and negative alternation of the supply voltage.

Capacitance (C): The capability to store charge in an electrostatic field. It can be expressed as equal to the charge Q in coulombs that is stored divided by the voltage E in volts that supplied the charge. Capacitance tends to oppose any change in voltage. The unit is farads.

Capacitive reactance (X_c): The opposition that a capacitor offers to a time changing signal or supplied voltage. Its value is $X_c = 1/2\pi fC$

Capacitor (C): A device made up of two metallic plates separated by a dielectric or insulating material. Used to store electrical energy in the electostatic field between the plates.

Cathode (K): The negative electrode of a semiconductor diode.

Charge (Q): A measurable quantity of electrical energy representing the electrostatic forces between atomic particles. Electrons have a negative charge.

Choke: An inductance which is designed to pass large amounts of dc current. It usually is used in power supply filters to help reduce ripple; although, there are inductances called rf chokes (rfc) which prevent rf from feeding to a circuit.

Circuit: A complete path that allows electrical current from one terminal of a voltage source to the other terminal.

Circuit breaker: An electromagnetic switch used as a protective device. It breaks a circuit if the current exceeds a specified value.

Clock or Clock generator: An electronic circuit that generates accurate and precisely controlled, regularly occurring synchronizing or timing signals called clock signals.

Clock rate: The frequency of oscillation of the master clock, or oscillator, in a system.

Coil: The component that is formed when several turns of wire are wound on a cylindrical form or on a metal core.

Collector (C): The element in a transistor that collects the moving electrons or holes, and from which the output usually is obtained. Analagous to the plate of a triode vacuum tube.

Color code: A system in which colors are used to identify the value of electronic components, or other variables, such a component tolerance.

Component: The individual parts that make up a circuit, a function, a subsystem or a total piece of equipment.

Conductor: A substance through which electrons flow with relative ease.

Contactor: A special relay for switching heavy currents at power line voltages.

Continuity: A continuous electrical path.

Controlled rectifier: A four-layer semiconductor device in which conduction is triggered ON by gate current and OFF by reducing the anode voltage below a critical value.

Coulomb (C): The unit of electrical charge, made up of a quantity of 6.25×10^{18} electrons.
Current (I): The flow of electrons, measured in amperes. One ampere results when one volt is impressed on a circuit that has a resistance of one ohm.
Decibel (dB): The standard unit for expressing the ratio between powers P_1 and P_2. $dB = 10 \log_{10} P_1/P_2$, one tenth of a bel.
Dielectric: The non-conducting material used to separate the plates of a capacitor or for insulating electric contacts.
Digital signal: A signal whose level has only discrete values, like on or off, 1 or 0, +5V or +0.2V.
Digital to Analog Conversion (or Converter) DAC or D/A): A circuit that converts a digital input signal to an analog output signal.
Diode: A device which has two terminals and has a high resistance to current in one direction and a low resistance to current in the other direction.
Direct Current (dc): Current in a circuit in one direction only.
Drain: The element in field-effect transistor which is roughly analagous to the collector of a bipolar transistor.
Effective value: The value of ac current that will produce the same heating effect in a load resistor as the corresponding value of dc current.
Electricity: A form of energy produced by the flow of electrons through materials and devices under the influence of an electromotive force produced electrostatically, mechanically, chemically or thermally.
Electrolytic capacitor: A capacitor whose electrodes are immersed in a wet electrolyte or dry paste.
Electromotive force (E): The force which causes an electrical current in a circuit when there is a difference in potential. Synonym for voltage.
Electron: The basic atomic particle having a negative charge that rotates around a positively charged nucleus of an atom.
Electrostatic field: The electrical field or force surrounding objects that have an electrical charge.
Emitter (E): The semiconductor material in a transistor that emits carriers into the base region when the emitter-base junction is forward biased.
Error: Any deviation of a computed, measured, or observed value from the correct value.
Farad (F): The basic unit for capacitance. A capacitor has a value of one farad when it has stored one coulomb of charge with one volt across it.
Field coil: An electromagnet formed from a coil of insulated wire wound around a soft iron core. Commonly used in motors and generators.
Field-Effect Transistor (FET): A 3-terminal semiconductor device where current is from source to drain due to a conducting channel formed by a voltage field between the gate and the source.
Filament: The heated element in an incandescent lamp or vacuum tube.
Filter: A circuit element or group of components which passes signals of certain frequencies while blocking signals of other frequencies.
Fluorescent: The ability to emit light when struck by electrons or other radiation.
Forward resistance: The resistance of a forward-biased junction when there is current through the semiconductor p-n junction.
Forward voltage (or bias): A voltage applied across a semiconductor junction in order to permit forward current through the junction and the device.
Frequency (F or f): The number of completed cycles of a periodic waveform during one second.
Gain (G): 1. Any increase in the current, voltage or power level of a signal. 2. The ratio of output to input signal level of an amplifier.
Ground (or Grounded): 1. The common return path for electric current in electronic equipment. Called electrical ground. 2. A reference point connected to, or assumed to be at zero potential with respect, to the earth.
Hall-Effect: A small voltage generated by a semiconductor carrying current when placed in a properly oriented magnetic field.
Henry (H or h): The unit of inductance. The inductance of a coil of wire in henries is a function of the coil's size, the number of turns of wire and the type of core material.
Hertz (Hz): One cycle per second.
Impedance (Z): In a circuit, the opposition that circuit elements present to alternating current. The impedance includes both resistance and reactance.
Inductance (L): The capability of a coil to store energy in a magnetic field surrounding it which results in a property that tends to oppose any change in the existing current in the coil.
Inductive reactance (X_L): The opposition that an inductance offers when there is an ac or pulsating dc in a circuit. $X_L = 2\pi fL$.
Input impedance: The impedance seen by a source when a device or circuit is connected across the source.
Integrated circuit (IC): A complex semiconductor structure that contains all the circuit components for a high functional density analog or digital circuit interconnected together on a single chip of silicon.
Junction: The region separating two layers in a semiconductor material; e.g., a p-n junction.
Junction transistor: A PNP or NPN transistor formed from three alternate regions of p and n type material. The alternate materials are formed by diffusion or ion implantation.
Leakage (or Leakage current): The undesired flow of electricity around or through a device or circuit. In the case of semiconductors, it is the current across a reverse-biased semiconductor junction.
Linear amplifier: A class A amplifier whose output signal is directly proportional to the input signal. The output is an exact reproduction of the input except for the increased gain.
Load: Any component, circuit, subsystem or system than consumes power delivered to it by a source of power.
Loop: A closed path around which there is a current or signal.
Magnetic Field: The force field surrounding a magnet.

Magnetic lines of force: The imaginary lines called flux lines used to indicate the directions of the magnetic forces in a magnetic field.
Megohm (MΩ): A million ohms. Sometimes abbreviated meg.
Microampere (μA): One millionth of an ampere.
Microfarad (μF): One millionth of a farad.
Milliampere (mA): One thousandth of an ampere.
Millihenry (mH): One thousandth of a henry.
Milliwatt (mW): One thousandth of a watt.
NPN Transistor: A bipolar transistor with a p-type base sandwiched between an n-type emitter, and an n-type collector.
N-type semiconductor material (N): A semiconductor material in which the majority carriers are electrons, and there is an excess of electrons over holes.
Ohm (Ω): The unit of electrical resistance. A circuit component has a resistance of one ohm when one volt applied to the component produces a current of one ampere.
Ohms-per-volt: The sensitivity rating for a voltmeter. Also expresses the impedance (resistance) presented to a circuit by the meter when a voltage measurement is made.
Open circuit: An incomplete path for current.
Operating point: The steady-state or no-signal operating point of a circuit or active device.
Operational amplifier (OP AMP): A high-gain analog amplifier with two inputs and one output.
Oscillation: A sustained condition of continuous operation where the circuit outputs a constant signal at a frequency determined by circuit constants and as a result of positive or regenerative feedback.
Pi (π): The mathematical constant which is equal to the ratio of the circumference of a circle to its diameter. Approximately 3.14.
Picofarad (pF): A unit of capacitance that is 1×10^{-12} farads or one millionth of a millionth of a farad.
PNP Transistor: A bipolar transistor with an n-type base sandwiched between a p-type emitter and a p-type collector.
Polarity: The description of whether a voltage is positive or negative with respect to some reference point.
Potential difference: The voltage difference between two points, calculated algebraically.
Power (P): The time rate of doing work.
Power (reactive): The product of the voltage and current in a reactive circuit measured in volt-amperes (apparent power).
Power (real): The power dissipated in the purely resistive components of a circuit measured in watts.
Power supply: A defined unit that is the source of electrical power for a device, circuit, subsystem or system.
P-type semiconductor material (P): A semiconductor material in which holes are the majority carriers and there is a deficiency of electrons.
Reactance (X): The opposition that a pure inductance or a pure capacitance provides to current in an ac circuit.
Rectification: The process of converting alternating current into pulsating direct current.
Relay: A device in which a set of contacts is opened or closed by a mechanical force supplied by turning on current in an electromagnet. The contacts are isolated from the electromagnet.
Resistance (R): A characteristic of a material that opposes the flow of electrons. It results in loss of energy in a circuit dissipated as heat.
Resistor (R): A circuit component that provides resistance to current in the circuit.
Reverse current: The current when a semiconductor junction is reverse biased.
Root-Mean-Square (RMS): See effective value. The RMS value of an ac sinusoidal waveform is 0.707 of the peak amplitude of the sine wave.
Semiconductor: One of the materials falling between metals as good conductors and insulators as poor conductors in the periodic chart of the elements.
Shunt: A parallel circuit branch, see Ammeter shunt.
Signal: In electronics, the information contained in electrical quantities of voltage or current that forms the input, timing, or output of a device, circuit, or system.
Silicon Controlled Rectifier (SCR): A semiconductor diode in which current through a third element, called the gate, controls turn-on, and the anode-to-cathode voltage controls turn-off.
Sine (sinusoidal) wave: A waveform whose amplitude at any time through a rotation of an angle from 0° to 360° is a function of the sine of an angle.
Step-down transformer: A transformer in which the secondary winding has fewer turns than the primary.
Step-up transformer: A transformer in which the secondary winding has more turns than the primary.
Transformer: A set of coils wound on an iron core in which a magnetic field couples energy between two or more coils or windings.
Transistor: A three-terminal semiconductor device used in circuits to amplify electrical signals or to perform as a switch to provide digital functions.
Turns ratio: The ratio of secondary winding turns to primary winding turns of a transformer.
Vector: A line representing the magnitude and time phase of some quantity, plotted on rectangular or polar coordinates.
Voltage (or Volt): The unit of electromotive force that caused current when included in a closed circuit. One volt causes a current of one ampere through a resistance of one ohm.
Voltage drop: The difference in potential between two points caused by a current through an impedance or resistance.
Watt (W): The unit of electrical power in joules per second, equal to the voltage drop (in volts) times the current (in amperes) in a resistive circuit.

Index

Accuracy
- Analog: 23
- D'Arsonval: 19, 20
- DMM: 23, 36
- Compromise: 46

Ammeter
- Analog: 12-14
- Digital: 25-28
- Clamp-on: 56, 104, 106, 131,137
- Multirange: 12-14
- Simple: 12-14

Amplifier: 87, 94, 95, 96
- Radio and TV: 95

Analog to digital conversion: 21-23, 33
Auto polarity: 38

Battery
- Current drain: 25-27, 77
- Testing: 4-6, 76-78, 129-132

Capacitor
- Automotive: 142
- Basics: 62, 63
- Defective: 92-94
- Electrolytic: 65
- Leakage current: 65
- Leakage resistance: 64
- Relative amount: 63

Ceiling fan: 127, 128
Chassis ground: 41
Circuit
- Parallel: 40, 41
- Series: 39, 40

Coffee maker: 113
Control circuits: 143-150
Current direction
- Conventional: 39
- Electron: 39

Current drain: 25-27, 130-132

Damping: 14
D'Arsonval meter movement: 1, 6, 20
Display
- Digital: 3, 22, 24, 34-37, 150
- Number of digits: 34-36

Dryer, clothes: 110-113
DMM
- Accuracy: 23, 36
- Basic: 21, 23, 33-35
- Characteristics: 21-23, 35
- Data hold: 38
- Display: 3, 22, 24, 34-37, 150
- Input impedance: 37
- Protection: 37, 52
- Resolution: 36
- Response time: 37

Electric lighting
- Fluorescent: 121
- Incandescent: 116, 117, 120, 121
- Low voltage: 122, 123

Fan, ceiling: 127, 128
Garage door opener: 125, 126
Grounding: 118, 119, 143
Ground reference: 41, 95

Hall-effect: 57, 58, 131, 137
Heating and air conditioning: 97-104

Ignition, engine: 138-142
Impedance: 53-56
- Input, of DMM: 37

Inductor: 151

Leakage test: 134
Lighting: 116, 120-123

Magnetic field: 20
Measurement
- AC: 53
- AC current: 27, 28, 55, 106
- AC voltage: 9, 25, 54, 55
- Amplifier: 95
- Automotive charging system: 129-138
- Automotive starter circuit: 41-42
- Bipolar junction transistor (BJT): 33, 71-73, 87-93
- Capacitance: 62-65
- Coffee maker: 113
- Continuity: 16, 30, 31
- Control circuits: 143-150
- Current in reactive circuit: 55, 58
- DC current (with clamp-on): 57, 131, 137
- DC current (with DMM): 25, 52
- DC current (with VOM): 12, 48
- DC power supply: 79-81
- DC voltage (with DMM): 24, 51
- DC voltage (with VOM): 4, 47
- Defective resistors: 92-94
- Defective transistors: 90, 91
- Diodes: 32, 70, 133, 134, 139, 147
- Doorbell: 81-83
- Dryer, clothes: 110-113
- Electrolytic capacitor: 65
- Error: 45, 46
- Fan, ceiling: 127
- Field-effect transistor (FET): 74, 75, 94
- Fixed resistor: 59, 60
- Gain of transistor: 90

Garage door opener: 125
Heating and air conditioning: 97-104
Ignition: 138-142
Impedance: 53-56
Lighting: 116, 120-123
Optical sensors: 124, 125
Outlet, ac: 10, 25, 26, 118, 119
Power, ac and dc: 58
Power distribution, ac: 116, 117
Power supply, dc: 79-81
Resistance (with DMM): 28, 52
Resistance (with VOM): 15, 50
SCR: 76
Semiconductor devices: 31-33, 69
Telephone: 83-87
Temperature: 104
Variable resistor: 60, 61
Voltage in reactive circuit: 54, 55
Voltage regulator: 129, 134, 135
Washer, clothes: 105-110
Meter
Analog: 1, 2
Clamp-on current: 56, 106, 137
D'Arsonval: 6, 20
Digital: 1-3
Loading: 45, 46
Protection: 19
Safety: 79
Multimeter: 1
Multitester
Analog: 1-4, 21, 23
Digital: 2, 3, 22, 23

Ohmmeter: 15-18
Batteries: 19, 50, 52
Ohms adjust control: 4, 15, 19
Ohm's law: 7, 39
Ohms per volt: 8
Optical sensor: 124, 125
Outlet, ac: 118, 119
Overrange: 36

Parallax: 3, 19
Parallel circuit: 40, 41
Photoconductor: 62, 124
Polarity
Defined: 4, 39, 41-43
DMM: 24, 38, 51
Of voltage drop: 43, 44
Power distribution, ac: 116, 117
Reactance: 53, 54
Rectifier
Bridge: 11
Alternator: 133-139
Reference point: 41, 95

Resistance
Internal, of meter: 7, 8
Measurement (with DMM): 28, 52
Measurement (with VOM): 15, 50
Multiplier: 7, 8
Resistor
Color code: 151
Fixed: 59
Measurement: 59
Shifty: 62
Thermally intermittent: 61
Variable: 60
Resolution of DMM: 36
Response time of DMM: 37
RMS: 37, 53

Schematic diagram: 7, 97
Semiconductor
Defective transistors: 90, 91
Diodes: 32, 70, 133, 134, 139, 147
SCR: 76
BJT: 33, 71-73, 87-93
FET: 74, 75, 94
Sensitivity, of meter: 8
Series circuit: 39, 40
Shunt: 12-14
Speed control circuits: 143-150

Temperature probe: 104
Test leads: 4
Thermistor: 62
Timers: 106-109
Transformer
Basics: 66-68
Continuity: 69
Shorts: 69
Step-down: 68
Step-up: 68
Turns ratio: 67

Varistor: 62
Vector addition: 54
Voltage regulator: 134-137
Voltmeter
AC: 11
Analog: 1-3, 21, 23
Basic: 7
Digital: 3, 22
Measurement: 4
Multirange ac: 11
Multirange dc: 8
VOM: 1-2, 46-47
Accuracy: 23
Protection: 19

Washer, clothes: 105-110